Insects and Diseases

by Rennie W. Doane

INSECTS AND DISEASE

CHAPTER I

PARASITISM AND DISEASE

PARASITES

The dictionary says that a parasite is a living organism, either animal or plant, that lives in or on some other organism from which it derives its nourishment for a whole or part of its existence. This definition will serve as well as any, as it seems to include all the forms that might be classed as parasites. As a general thing, however, we are accustomed to think of a parasite as working more or less injury to its host, or perhaps we had better say that if it does not cause any irritation or ill effects its presence is not noted and we do not think of it at all.

As a matter of fact the number of parasitic organisms that are actually detrimental to the welfare of their hosts is comparatively small while the number of forms both large and small that lead parasitic lives in or on various hosts, usually doing no appreciable harm, often perhaps without the host being aware of their presence, is very great indeed.

Few of the higher animals live parasitic lives. The nearest approach to a true parasite among the vertebrates is the lamprey-eel (Fig. 1) which attaches itself to the body of a fish and sucks the blood or eats the flesh. Among the Crustaceans, the group that includes the lobsters and crabs, we find many examples of parasites, the most extraordinary of which is the curious crab known as Sacculina (Fig. 2). In its early stages this creature is free-swimming and looks not unlike other young crabs. But it soon attaches itself to another crab and begins to live at the expense of its host. Then it commences to undergo remarkable changes and finally becomes a mere sac-like organ with a number of long slender root-like processes penetrating and taking nourishment from the body of the unfortunate crab-host.

The worms furnish many well-known examples of parasites, whole groups of them being especially adapted to parasitic life. The tapeworms, common in many animals and often occurring in man, the roundworms of which the trichina (Fig. 3) that causes "measly" pork is a representative, are familiar examples. These and a host of others all show a very high degree of specialization fitting them for their peculiar lives in their hosts.

From among the insects may be selected interesting examples of almost all kinds and degrees of parasitism, temporary, permanent, external, internal (Figs. 4, 5, 6). Among them is found, too, that curious condition known as hyperparasitism where one animal, itself a parasite, is preyed upon by a still smaller parasite.

"The larger fleas have smaller fleas Upon their backs to bite um, These little fleas still smaller fleas And so ad infinitum."

Coming now to the minute, microscopic, one-celled animals, the Protozoa, we find entire groups of them that are living parasitic lives, depending wholly on one or more hosts for their existence. Many of these have a very remarkable life-history, living part of the time in one host, part in another. The malarial parasite and others that cause some of the diseases of man and domestic animals are among the most important of these.

PARASITISM

Among all these parasites, from the highest to the lowest the process that has fitted them for a parasitic life has been one of degeneration. While they may be specialized to an extreme degree in one direction they are usually found to have some of the parts or organs, which in closely related forms are well developed, atrophied or entirely wanting. As a rule this is a distinct advantage rather than a disadvantage to the parasite, for those parts or organs that are lost would be useless or even in the way in its special mode of life.

Then, too, the parasite often gives up all its independence and becomes wholly dependent on its host or hosts not only for its food but for its dissemination from one animal to another, in order that the species may not perish with the host. But in return for all this it has gained a life of ease, free from most of the dangers that beset the more independent animals, and is thus able to devote its whole time and energy to development and the propagation of the species.

We are accustomed to group the parasites that we know into two classes, as harmful or injurious and as harmless, the latter including all those kinds that do not ordinarily affect our well-being in any way. But such a classification is not always satisfactory or safe, for certain organisms that to-day or under present conditions are not harmful may, on account of a great increase in numbers or change of conditions, become of prime importance to-morrow. An animal that is well and strong may harbor large numbers of parasites which are living at the expense of some of the host's food or energy or comfort, yet the loss is so small that it is not noticed and the intruders, if they are thought of at all, are classed as harmless. Or we may at times even look upon them as beneficial in one way or another. "A reasonable amount of fleas is good for a dog. They keep him from brooding on being a dog."

But should these parasites for some reason or other increase rapidly they might work great harm to the host. Even David Harum would limit the number of fleas on a dog. Or the animal might become weakened from some cause so that the drain on its resources made by the parasites, even though they did not increase in numbers, would materially affect it.

Perhaps the most serious way in which parasites that are usually harmless may become of great importance is illustrated by their introduction into new regions or, as is more often the case, by the introduction of new hosts into the region where the parasites are found. Under normal conditions the animals of a given region are usually immune to the parasites of the same region. That is, they actually repel them and do not suffer them to exist in or on their bodies, or they are tolerant toward them. In the latter case the parasites live at the expense of the host, but the host has become used to their being there, adapted to them, and the injury that they do, if any, is negligible.

But when a new animal comes into the region from some other locality the parasites may be extremely dangerous to it. There are many striking examples of this. Most of the people living in what is known as the yellow fever belt are immune to the fever. They will not develop it even under conditions that would surely mean infection for a person from outside this zone. Certain of our common diseases which we regard as of little consequence become very serious matters when introduced among a people that has never known them before. The cattle of the southern states are immune to the Texas fever, but let northern cattle be sent south or let the ticks which transmit the disease be taken north where they can get on cattle there, and the results are disastrous.

Another striking example and one that is attracting world-wide attention just now is the trypanosome that is causing such devastation among the inhabitants of central Africa. With the advent of white men into this region and the consequent migration of the natives along the trade routes this parasite, which is the cause of sleeping sickness, is being introduced into new regions and thousands upon thousands of people are dying as a result of its ravages.

DISEASES CAUSED BY PARASITES

Some two hundred years ago, after it became known that minute animal parasites were associated with certain diseases and were the cause of them, it rapidly came to be believed that all our ills were in some way caused by such parasites, known or unknown. Further study and investigation failed to reveal the intruders in many instances and so it began to be doubted whether after all they were responsible for much that had been laid at their doors. Then after it was discovered that minute plant parasites, bacteria, were responsible for many diseases they in turn began to be accused of being the cause of most of the ills that the flesh is heir to.

In later years we have come to adopt what seems to be a more reasonable view, for we can see and definitely prove that neither of these extreme views was correct but that there was much truth in each of them. To-day we recognize that certain diseases, such as typhoid, cholera, tuberculosis and many others, are caused by the presence of bacteria in the body, and it is just as definitely known that such maladies as malaria and sleeping sickness are caused by animal parasites.

Then there is a long list of other epidemic diseases, such as smallpox, measles and scarlet fever, the exact cause of which has not been determined. Many of these are believed to be due to micro-organisms of some kind, and if so they will almost certainly sooner or later be found. Curiously enough most of the diseases in this last class and many of those in the first are contagious, while all that are caused by animal parasites are, as far as is known, infectious but not contagious.

INFECTIOUS AND CONTAGIOUS DISEASES

It is important that we keep in mind this distinction. By contagious diseases are meant those that are transmitted by contact with the diseased person either directly, by touch, or indirectly by the use of the same articles, by the breath or effluvial emanations from the body or other sources. Small-pox, measles, influenza, etc., are examples of this group. By infectious diseases are meant those which are disseminated indirectly, that is, in a roundabout way by means of water or food or other substances taken into or introduced into the body in some way. Typhoid, malaria, and yellow fever, cholera and others are examples of this class. Thus it is evident that all of the contagious diseases may be infectious, but many of the infectious diseases are not as a rule contagious, although some of them may become so under favorable conditions.

Just one example will show the importance of knowing whether a disease is contagious or infectious. Until a few years ago it was believed that yellow fever was highly contagious and every precaution was taken to keep the disease from spreading by keeping the infected region in strict quarantine. This often meant much hardship and suffering and always a great financial loss. We now know that it is infectious only and not contagious, and that all this quarantine was unnecessary. The whole fight in controlling an outbreak of yellow fever or in preventing such an outbreak is now directed against the mosquito, the sole agent by which the disease can be transmitted from one person to another.

EFFECT OF THE PARASITE ON THE HOST

We have seen how a few parasites in or on an animal do not as a rule produce any appreciable ill effects. This is of course a most fortunate thing for us, for the parasitic germs are everywhere.

There is perhaps "more truth than poetry" in the following newspaper jingle:

"Sing a song of microbes, Dainty little things, Eyes and ears and horns and tails, Claws and fangs and stings. Microbes in the carpet, Microbes in the wall, Microbes in the vestibule, Microbes in the hall. Microbes on my money, Microbes in my hair, Microbes on my meat and bread, Microbes everywhere. Microbes in the butter, Microbes in the cheese, Microbes on the knives and forks, Microbes in the breeze. Friends are little microbes, Enemies are big, Life among the microbes is-- Nothing 'infra dig.' Fussy little microbes, Millions at a birth, Make our flesh and blood and bones, Keep us on the earth."

While of course most of these microbes are to be regarded as absolutely harmless and some as very useful, many have the power to do much injury if the proper conditions for their rapid development should at any time exist. While the size of the parasite is always a factor in the damage that it may do to the host the factor of numbers is perhaps of still greater importance because of the power of very rapid multiplication possessed by so many of the smaller forms.

Certain minute parasites in the blood may cause little or no inconvenience, but should they begin to multiply too rapidly some of the capillaries may be filled up and trouble thus result. Or take some of the larger forms. A few intestinal worms may cause no appreciable effect on the host, but as soon as their numbers increase serious conditions may come about simply by the presence of the great masses in the host even if they were not robbing it of its nourishment. Many instances are known where such worms have formed masses that completely clogged up the alimentary canal. Such injuries as these may be regarded as mechanical injuries. Some parasites injure the host only when they are laying their eggs or reproducing the young. These may clog up passages or some of them may be carried to some more sensitive part of the body where the damage is done. The guinea-worm of southwestern Asia and of Africa lives in the body of its host for nearly a year sometimes attaining a great length and migrating through the connective tissue to different parts of the body causing no particular inconvenience until it is ready to lay its eggs when it comes to the surface and then great suffering may result. The African eye-worm is another example of a parasite causing mechanical injury only at certain times. It works in the tissues of the body sometimes for a long while, doing no harm unless it finds its way to the connective tissue of the eyeball.

It is known that many of the germs which cause diseases cannot get into the body unless the protecting membranes have first been injured in some way. Thus the germs that cause plague and lockjaw find their way into the system principally through abrasions of the skin. Many physicians have come to believe that the typhoid fever germ cannot get into the body from the intestine where it is taken with our food or drink unless the walls of the intestine have been injured in some way. It is well known that of the many parasites that inhabit the alimentary canal some rasp the surface and others bore through into the body cavity. This in itself may not be a serious thing, but if the mechanical injury thus caused opens the way for malignant germs, baneful results may follow. Even that popular disease appendicitis is believed to be due sometimes to the injury caused by the work of parasites in the appendix.

Parasites may cause morphological or structural changes in the tissues of their hosts. The stimulation caused by their presence may result in swellings or excrescences or other abnormal growths. Interesting examples of this are to be found in the way in which pearls are formed in various mollusks. In the pearl oysters of Ceylon occur some of the best pearls. If these are carefully sectioned there may usually be found at the center the remains of certain cestode larvaewhose presence in the oyster caused it to deposit the nacreous layers that make up the pearl. Other parasites cause similar growths in various shellfish. The great enlargements of the arms or legs or other parts of the body seen in patients affected with elephantiasis is an abnormal growth due to the presence of the parasitic filar?in some of the lymph-glands where they have come to rest.

Finally, the parasite may exert a direct physiological effect on the host. This is evident when the parasite demands and takes a portion of the nourishment that would otherwise go to the building up of the host. Sometimes this is of little importance, but at other times it may be a matter of life or death to the infected animal. The physiological effect produced may be due to the toxins or poisonous matters that are given off by the parasite while it is living in the host's body. Thus it is believed that the malarial patients usually suffer less from the actual loss of red blood-corpuscles that are destroyed by the parasite than they do from the effects of the poisonous excretions that are poured into the circulation when the thousands of corpuscles break to release the parasites.

One other point in regard to the relation of the parasite to its host and this part of the subject may be dismissed. From our standpoint we look upon the presence of parasites in the body as an abnormal condition. From a biological standpoint their presence there is perfectly normal; it is a necessary part of their life. We think that they have no business there, but from the viewpoint of the parasites their whole business is to be just there. If they are not, they perish. And when we take a dose of quinine or other drug we are killing or driving from their homes millions of these little creatures who have taken up their abode with us for the time being. But they interfere with our health and comfort, so they must go.

CHAPTER II

BACTERIA AND PROTOZOA

BACTERIA

On the border line between the plant and the animal worlds are many forms which possess some of the characteristics of both. Indeed when an attempt is made to separate all known organisms into two groups one is immediately confronted with difficulties. In looking over the text-books of Botany we will find that certain low forms are discussed there as belonging with the plants, and on turning to the manuals of Zoology we will find that the same organisms are placed among the lowest forms of animals. The question is of course of little actual importance from certain points of view. It serves, however, to show the close relation of all forms of life, and from a medical standpoint it may be of very great importance owing to the difference in the life-habits, methods of reproduction and methods of transmission of many of the forms that cause disease. We have already seen that none of the diseases that are caused by animal parasites is contagious, while many of those caused by bacteria are both contagious and infectious.

Just over on the plant side of this indefinite border line are the minute organisms known as bacteria. Their numbers are infinite and they are found everywhere. The majority of them are beneficial to mankind in one way or another, but some of them cause certain of the diseases that we will have to discuss later so attention may be called here to a few of the important facts in regard to their organization and life-history in order that we may better understand how they may be so easily transferred from one host to another.

Although these bacilli are so extremely minute (Fig. 7), some of them so small that they cannot be seen with the most powerful microscopes, they differ in size, shape, methods of division and spore-formation. Each species makes a characteristic growth on gelatin, agar or other media upon which it may be cultivated. In this way as well as by the inoculation of animals the presence of the ultramicroscopic kinds may be demonstrated.

The method of reproduction is very simple. They increase to a certain point in size, then divide. This growth and division takes place very rapidly. Twenty to thirty minutes is sufficient time in some cases for a just-divided cell to attain full size and divide again. Thus in a few days time the number of bacteria resulting from a single individual would be inconceivable if they should all develop.

Fortunately for us, however, they do not all multiply so rapidly as this and besides there are natural checks, not the least of which are the substances given off by the bacteria themselves in their growth and development. Such excretions often serve to inhibit further multiplication. Sometimes, though not often, they form spores which not only provide for a more rapid multiplication, but enable the organism to live under conditions that would otherwise prove fatal to it.

Bacteria may be conveniently grouped under two heads: those that live upon dead organic matter, known as the saprophytic forms, and those that are found in living plants or animals, the true parasites. Such a grouping is not always entirely satisfactory, for many of the kinds that live saprophytically under normal conditions may become parasitic if opportunity offers, and also many of those that are usually regarded as parasitic may be grown in cultures of agar or other media, under which conditions they may be regarded as living saprophytically.

It is this power of easily adapting themselves to different conditions that makes many of the kinds dangerous. The bacillus which causes tetanus or lockjaw will illustrate this. It is a rather common bacillus in soil in many localities. As long as it remains there it is of no special importance, but if it is introduced into the body through a scratch or any other wound it becomes a very serious matter.

We may say, then, that the effect the bacillus has on the host depends largely on the host. Not only does it depend on what the host is, but the particular condition of the host at the time of infection is of importance. Children are subject to many diseases that adults seldom have. Hunger, thirst, fatigue, exposure and other factors may make a person susceptible to the actions of certain bacteria that would be harmless under other conditions.

The minute size and great numbers of the bacteria make their dissemination a comparatively simple matter. They may be carried in the air as minute particles of dust; they may be carried in water or milk; they may be carried on the clothing or on the person from one host to another, or they may be disseminated in scores of other ways. In other chapters, particularly the one dealing with the house-fly and typhoid, we shall see how it is that insects are often important factors in spreading some of the most dreaded of the bacterial diseases.

THE PROTOZOA

The Protozoa, or one-celled animals, belonged to an unknown world before the invention of the microscope. The first of these instruments enabled the early observers to see some of the larger and more conspicuous members of the group and each improvement of the microscope has enabled us to see more and more of them and to study in detail not only the structure but to follow the life-history of many of them.

The Amoeba. With some, as the common amoeba (Fig. 8), a minute little form that is to be found in the slime at the bottom of almost any body of water, the life-history is extremely simple. The organism itself consists of a minute particle of protoplasm, a single cell with no definite shape or body-wall and no specialized organs or apparatus for carrying on the life-functions. It lives in the slime or ooze in fresh or salt water, takes its food by simply flowing over the particle that is to be ingested, grows to a certain limit of size, then divides into two more or less equal parts, each part becoming a new animal that goes on with its development as did the parent form. This process of growth and division may go on for many generations, but cannot continue indefinitely unless there is a conjugation of two separate individuals. This process of conjugation is just the opposite to that of division. Two amoeba flow together and become one. It seems to rejuvenate the organism so that it is able to go on with its division and thus fulfil its life-mission which is the same for these lowly animals as with the higher, that of perpetuating the species.

Classes of Protozoa. The group or Phylum Protozoa is divided into four smaller groups or classes. The amoeba belongs to the lowest of these, the Rhizopoda. Rhizopoda means "root-footed," and the name is applied to these animals because most of them move about by means of root-like processes known as pseudopodia or "false feet." This is by far the largest class and contains thousands of forms, mostly living in salt water but there are many fresh-water species. They are non-parasitic, but some of them by their presence in the body may cause such diseases as dysentery, etc.

The next class which may be known as the whip-bearers (Mastigophora) includes those Protozoa that move by fine undulating processes called flagella. One of the common representatives of this class is the little green Euglena (Fig. 9), whose presence in standing ponds and puddles often imparts a greenish color to the water. Then in the salt water near the surface there are often myriads of minute Noctiluca whose wonderfully phosphorescent little bodies glow like coals of fire when the water is disturbed at night. Although this class contains fewer forms than the preceding some of these have within recent years been found to be of great importance because they live as parasites on man and other animals. The trypanosome whose presence in the blood and tissues of the patient causes that dreadful disease which ends in sleeping sickness belongs here as well as do several other similar kinds that produce serious troubles for various mammals and birds. The Spirocha, about which there has been so much recent discussion, also belong here. These are simple spiral-like forms (Fig. 10), that are sometimes classed with the simple plants, bacteria, but Nuttall and others have shown very definitely that they should be classed with the simplest animals, the Protozoans. These are the cause of relapsing fevers in man and of several diseases of domestic animals. It is believed by certain eminent zoologists that when the germ that causes yellow fever is discovered it will be found to belong to this group.

The members of the class Infusoria, so called because they were early found to be abundant in various infusions, are characterized by numerous fine cilia or hair-like organs by means of which the organism moves about and procures its food. The well-known "slipper animalcule" (Paramoecium) (Fig. 11), and the "bell-animalcule" (Vorticella) (Fig. 12) are two common representatives. The Paramoecia were the animals mostly used by Jennings in his wonderfully interesting experiments on the behavior of these lowly forms of life. He showed that they always reacted in a certain definite way in response to particular stimuli, and he was led to believe that he could see "what must be considered the beginnings of intelligence and of many other qualities found in the higher animals." A species of Vorticella was probably the first Protozoan that was ever observed. An old Dutch microscopist, Anton von Leeuwenhoek, in 1675, while studying with lenses of his own manufacture, discovered and described forms which undoubtedly belong to this genus. Few if any of the Infusoria are pathogenic, although some are said to be associated with certain intestinal diseases both in man and the lower animals (Fig. 13).

The last class, the Sporozoa, or the spore-forming animals, while small in the number of known species, only about three hundred kinds being known, is extremely important. A number of diseases in man and other animals are due to the presence of these Sporozoans, for they are all parasitic. Few if any animals are exempt from their attacks. They even attack other minute Protozoa. One hundred and fifty-seven species have been recorded as attacking insects, one hundred species attack birds, fifty-two reptiles, eighty crustaceans, twenty-two fish, and so through the list. Ten have been recorded as attacking man. In some instances the parasite is always present in the host and some hosts may harbor several different species of Sporozoa.

Very little work had been done on this group of parasites prior to 1900. Since that time most of the species that we now know have been discovered, and within the last few years the life-histories of many of these have been worked out quite completely. No other group of animals is being studied more to-day by both the physicians and biologists.

The Sporozoa vary greatly in appearance, organization and life-history. They are so very plastic that they can adapt themselves readily to their various hosts, hence we have a great variety of forms. But they all agree in certain characters; all take their food and oxygen and carry on excretory processes by osmosis, i.e., through the body-wall; all are capable of some kind of locomotion, some have one or more flagella, others move by a pseudopod movement. Some are capable of moving from cell to cell in the body as do the white blood-corpuscles. They all agree in the production of spores--hence the name.

At certain stages in their development the nucleus within the body of the organism divides again and again until there are a great many daughter nuclei, each accompanied by a small mass of protoplasm, often inclosed in a little sac or cyst of its own. This is the process of spore-formation and we see that from a single individual we may have by division, not two animals as in the amoeba, but a score or more of them. The little cysts or capsules that inclose them enable them to resist without injury many vicissitudes that would otherwise destroy them. They may dry up or freeze or lie for a long time in the ground or water until the time comes when they are introduced into another host.

The class Sporozoa is divided into five small groups or orders. Nearly all of these contain forms that are of more or less importance, but the ones that live in the blood-cells (H 鎚 osporidiida) are of the most interest to us because the parasites that cause the malarial fevers and various other diseases belong here. These are dependent on two hosts for their existence, the sexual generation usually occuring in an insect or other invertebrate and the asexual generation in some vertebrate.

CHAPTER III

TICKS AND MITES

The other group or Phylum of animals with which we will be particularly concerned is known as the Arthropoda, which means "jointed-feet" and includes the crayfish, crabs, spiders, mites, ticks and insects. Of these only the last three are of interest to us now. It is customary to speak of spiders, mites and ticks as insects, but as they have four pairs of legs, instead of three pairs, in the adult stage, and as their bodies are not divided into three distinct regions as in the insects, they are placed in a different class.

GENERAL CHARACTERS OF TICKS

The ticks are all comparatively large, that is, they are all large enough to be seen with the unaided eye even in their younger stages and some grow to be half an inch long. When filled with blood the tough leathery skin becomes much distended often making the creature look more like a large seed than anything else (Fig. 14). This resemblance is responsible for some of the popular names, such as "castor-bean tick," etc.

The legs of most species are comparatively short, and the head is small so that they are often hardly noticeable when the body is distended. The sucking beak which is thrust into the host when the tick is feeding is furnished with many strong recurved teeth which hold on so firmly that when one attempts to pull the tick away the head is often torn from the body and left in the skin. Unless care is taken to remove this, serious sores often result.

Ticks are wholly parasitic in their habits. Some of them live on their host practically all their lives, dropping to the ground to deposit their eggs when fully mature. Others leave their host twice to molt in or on the ground. The female lays her eggs, 1,000 to 10,000 of them, on the ground or just beneath the surface. The young "seed-ticks" that hatch from these in a few days soon crawl up on some near-by blade of grass or on a bush or shrub and wait quietly and patiently until some animal comes along. If the animal comes close enough they leave the grass or other support and cling to their new-found host and are soon taking their first meal. Of course thousands of them are disappointed and starve before their host appears, but as they are able to live for a remarkably long time without taking food their patience is often rewarded and the long fast ended.

Those species which drop to the ground to molt must again climb to some favorable point and wait for another host on which they may feed for a while. Then they drop to the ground for a second molt and if they are successful in gaining a new host for the third time they feed and develop until fully mature and the female is ready to lay her eggs. The Texas fever tick, and some others, as we shall see, do not drop to the ground to molt but once having gained a host remain on it until ready to deposit their eggs.

The young ticks have only six legs (Fig. 15) but after the first molt they all have eight.

TICKS AND DISEASE

Texas Fever. Ever since stockmen began driving southern cattle into states further north it has been noted that the roads over which they were driven became a source of great danger to northern cattle. Often 80% to 90% of the native cattle died after a herd of southern cattle passed through their region and the losses became so great that both state and national laws were passed prohibiting the driving or shipping of southern cattle into northern states.

But for years the cause of this fever, which came to be known as the Texas fever, was not known. The southern cattle themselves seemed healthy enough and it was difficult to understand how they could give the disease to the others. It was early noticed, too, that it was not necessary for the northern cattle to come in direct contact with the others in order to contract the disease. Indeed the disease was not contracted in this way at all. All that was necessary for them was to pass along the same roads or feed in the same pastures or ranges. Still more puzzling was the fact that these places did not seem to become a source of danger until some weeks after the southern cattle had passed over them and then they might remain dangerous for months.

In 1886 Dr. Theobald Smith of the Bureau of Animal Industry, United States Department of Agriculture, found that the fever was caused by the presence in the infected cattle of a minute Sporozoan parasite (Piroplasma bigeminum). Further investigations and experiments proved conclusively that this parasite was transmitted from the infected to the well animal only by the common cattle tick now known as the Texas fever tick (Fig. 16).

The infection is not direct, that is, the tick does not feed on one host then pass to another carrying the disease germs with it. Unlike many other ticks the Texas fever tick does not leave its host until it is fully developed. When the female is full grown and gorged she drops to the ground and lays from 2,000 to 4,000 eggs which soon hatch into the minute "seed-ticks" which make their way to the nearest blade of grass or weed or shrub and patiently wait for the cattle to come along.

If the mother tick had been feeding on an animal that was infected with the

Texas fever parasite, her body was filled with the minute organisms of which some found their way into the eggs so that the young ticks hatching from them were already infected and ready to carry the infection to the first animal they fed upon.

It took many years of hard patient work to learn all this, but the knowledge thus obtained cleared up much of the mystery in connection with the occurrence of the fever in the north and, as we shall see, suggested the possibility of other diseases being communicated in the same way.

It was found that the southern cattle in the region where the ticks occur normally, usually have a mild attack of the disease when they are young and although they may be infected with the parasite all the rest of their lives it does not affect them seriously. These cattle are almost always infected with ticks, and when taken north where the ticks do not occur naturally and where the cattle are therefore non-immune, some of the mature ticks drop to the ground and lay their eggs which in a few weeks hatch out and are ready to infect any animal that passes by. The northern cattle not being used to the disease soon sicken and die.

It is estimated that the annual loss due to this disease and the ravages of the tick in the United States is over $100,000,000, so of course most determined efforts are being made to stamp it out. Formerly various devices for dipping the tick-infested cattle into some solution that would kill the ticks were resorted to, but it was always expensive and never very satisfactory. The immunizing of the cattle by inoculating them when they were young with infected blood has been practised. Very recent investigations have shown that it is possible and practicable to rid pastures of ticks by a system of feed-lots and pasture rotation. The aim is to have as many of the ticks as possible drop to the ground on land where they may be destroyed and to so regulate the use of the pasture that the ticks in all of them may eventually be left to starve.

Several similar diseases of cattle, many of them probably identical with Texas fever, occur in other parts of the world where the losses are sometimes appalling. Horses, sheep, dogs, and other animals are also affected with diseases caused by the same group of Protozoan parasites. Most of them have been shown to be transmitted by various species of ticks (Fig. 17) so that from an economical standpoint these little pests are becoming of prime importance. Not only do they transmit the disease germs that infect domestic animals but they are known to be responsible for at least two diseases of men, Rocky Mountain spotted fever and the relapsing fevers.

Spotted Fever. The first of these is a disease that for some years has been puzzling the physicians in Idaho and Montana and other mountainous states. A few years ago certain observers recorded finding Protozoan parasites in the blood of those suffering from the disease, and although more recent investigations have failed to confirm these particular observations it is now quite generally believed that the disease is caused by some such parasite and that the organism is transferred from one host to another by certain species of ticks that live on wild mammals of the region where the disease exists. Dr. H.T. Ricketts, who has made a special study of the disease, has shown:

"1. That the period of activity of the disease is limited to the season during which the adult female and male ticks attack man.

"2. That in practically all cases of this disease it can be shown that the patient has been bitten by a tick.

"3. That the period between the tick bite and the onset of the disease in the many animals he has experimented with corresponds very closely to this period as observed in man.

"4. That infected ticks are to be found in the locality where the disease occurs.

"5. That the virus of spotted fever is very intimately associated with the tissues of the tick's body as is shown by the fact that the female passes the infection on to her young through her eggs, and further, by the observation that in either of the two earlier stages of the life cycle the disease may be contracted by biting a sick animal and communicated to other animals after molting or even after passing through an intermediate stage."

Professor R.A. Cooley of Montana, from whose report the above quotation is taken, has also made studies of the habits of the tick and believes there can be no doubt that it is the disseminator of the disease.

Relapsing Fever. The relapsing fever is an infectious disease or possibly a group of closely related infectious diseases occurring in various parts of the world. Occasionally it is introduced into America, but it does not seem to spread here. It has been shown that the disease is communicated from one person to another by means of blood-sucking insects. In Central Africa where the disease is very prevalent a certain common tick (Ornithodoros moubata) (Fig. 18) is known to transmit the disease. This tick lives in the resting places and around the huts of the natives and has habits very similar to the bedbug of other climes, feeding at night and hiding during the day. It attacks both man and beast and is one of the most dreaded of all the African pests.

Nathan Bank, our foremost authority on ticks, in summing up the evidence against them says:

"It is therefore evident that all ticks are potentially dangerous. Any tick now commonly infesting some wild animal, may, as its natural host becomes more uncommon, attach itself to some domestic animal. Since most of the hosts of ticks have some blood-parasites, the ticks by changing the host may transplant the blood-parasites into the new host producing, under suitable conditions, some disease. Numerous investigators throughout the world are studying this phase of tick-life, and many discoveries will doubtless signalize the coming years."

MITES

The mites are closely related to the ticks, and although none of them has yet been shown to be responsible for the spread of any disease their habits are such that it would be entirely possible for some to transmit certain diseases from one host to another, from animal to animal, from animal to man, or from man to man. A number of these mites produce certain serious diseases among various domestic animals and a few are responsible for certain diseases of men.

Face-mites. Living in the sweat-glands at the roots of hairs and in diseased follicles in the skin of man and some domestic animals are curious little parasites that look as much like worms as mites (Fig. 19). Such diseased follicles become filled with fatty matter, the upper end becomes hard and black and in man are known as blackheads. If one of these blackheads is forced out and the fatty substance dissolved with ether the mites may be found in all stages of development. The young have six legs, the adult eight. The body is elongated and transversely wrinkled. In man they are usually found about the nose and chin and neck where they do no particular harm except to mar the appearance of the host and to indicate that his skin has not had the care it should have. Very recently certain investigators have found that the lepr?bacilli are often closely associated with these face mites and believe that they may possibly aid in the dissemination of leprosy. It is also thought that they may sometimes be the cause of cancer, but as yet these theories have not been proven by any conclusive experiment.

In dogs and cats these same or very similar parasites cause great suffering. In bad cases the hair falls out and the skin becomes scabby. Horses, cattle and sheep are also attacked. The disease caused by these mites on domestic animals is not usually considered curable except in its very early stages when salves or ointments may help some.

Itch-mites. "As slow as the seven-years' itch" is an expression, the meaning of which many could appreciate from personal experience, for it certainly seemed to take no end of time to get rid of the itch once it was contracted. Just why seven years should have been set for the limit of the disease is not clear, for if the little roundish mites that cause the disease live for seven years on a host they are not going to move out voluntarily even if their seven-year lease has expired.

The minute whitish mites (Fig. 20) that cause this disgusting disease are barely visible to the naked eye. They are usually very sluggish but become more active when warmed. They live in burrows just beneath the outer layer of skin, sometimes extending deeper and causing most intense itching. As the female burrows, she lays her eggs from which come the young mites that are to spread the infection. Various sulphur ointments and washes are used as remedies. Cleanliness will prevent infection.

Closely related to the itch-mite of man (Sarcoptes scabiei) are several kinds attacking domestic animals, causing mange, scab, etc. The variety infesting horses burrows in the skin and produces sores and scabs, and is a source of very great annoyance. These mites may also migrate to man. Tobacco water and sulphur ointments are used as remedies.

Horses and cattle are also infested by other mites (Psoroptes communis) which cause the common mange. These do not burrow into the skin but live outside in colonies, feeding on the skin and causing crusts or scabs. The inflammation causes the animal to scratch and rub constantly and often causes the loss of much of the hair.

Harvest-mites. A score or more of different varieties of mites cause many other diseases of domestic animals, such as the scab of sheep and hogs and chickens, various other manges of the horses and cattle and dogs, etc. But we need to call attention to just one more example, that of the harvest-mites or jiggers (Fig. 21). Professor Otto Lugger, from whose report on the Parasites of Man and Domestic Animals most of these notes in regard to the mites are taken, thus feelingly refers to this pest.

"About the very worst pests of man and domesticated animals are the Harvest-bugs, Red-bugs or Jiggers.... Men and animals passing through low herbage that harbors them are attacked by these pests, which, whenever they succeed in finding a host, burrow in and under the skin, causing intolerable itching and sores, the latter caused by the feverish activity of the finger-nails of the host, if that should be a man, whose energy in scratching, apparently, cannot be controlled and who is bound forcibly to remove the intruders. The writer has been there! Those who have ever passed through meadows infested with red-bugs will remember the occasion."

Horses, cattle, dogs and cats and other animals suffer also. Again sulphur ointments are the best remedies.

"The normal food of these mites must, apparently, consist of the juices of plants, and the love of blood proves ruinous to those individuals which get a chance to indulge it. For, unlike the true chigoe, the female of which deposits eggs in the wound she makes, these harvest-mites have no object of the kind, and when not killed at the hands of those they torment they soon die victims to their sanguinary appetite."

CHAPTER IV

HOW INSECTS CAUSE OR CARRY DISEASE

It has been estimated that there are about four thousand species or kinds of Protozoans, about twenty-five thousand species of Mollusks, about ten thousand species of birds, about three thousand five hundred species of mammals, and from two hundred thousand to one million species of insects, or from two to five times as many kinds of insects as all other animals combined.

Not only do the insects preponderate in number of species, but the number of individuals belonging to many of the species is absolutely beyond our comprehension. Try to count the number of little green aphis on a single infested rose-bush, or on a cabbage plant; guess at the number of mosquitoes issuing each day from a good breeding-pond; estimate the number of scale insects on a single square inch of a tree badly infested with San Jos?scale; then try to think how many more bushes or trees or ponds may be breeding their millions just as these and you will only begin to comprehend the meaning of this statement.

As long as these myriads of insects keep in what we are pleased to call their proper place we care not for their numbers and think little of them except as some student points out some wonderful thing about their structure, life-history or adaptations. But since the dawn of history we find accounts to show that insects have not always kept to their proper sphere but have insisted at various times and in various ways in interfering with man's plans and wishes, and on account of their excessive numbers the results have often been most disastrous.

Insects cause an annual loss to the people of the United States of over $1,000,000,000. Grain fields are devastated; orchards and gardens are destroyed or seriously affected; forests are made waste places and in scores of other ways these little pests which do not keep in their proper places are exacting this tremendous tax from our people.

These things have been known and recognized for centuries, and scores of volumes have been written about the insects and their ways and of methods of combating them.

But it is only in recent years that we have begun to realize the really important part that insects play in relation to the health of the people with whom they are associated. Dr. Howard estimates that the annual death rate in the United States from malaria is about twelve thousand, entailing an annual monetary loss of about $100,000,000, to say nothing of the suffering and misery endured by the afflicted. All this on account of two or three species of insects belonging to the mosquito genus Anopheles.

Yellow fever, while not so widespread, is more fatal and therefore more terrorizing. Its presence and spread are due entirely to a single species of mosquito. Flies, fleas, bedbugs, and many other insects have been shown to be intimately connected with the spread of several other most dreaded diseases, so it is no wonder that physicians, entomologists and biologists are studying with utmost zeal many of these forms that bear such a close relation not only to our welfare and comfort but to our lives as well.

It would be out of place to try to give here even a brief outline of the classification of insects, such as may be found in almost any of the many books devoted to their study.

The most generally accepted classification divides the insects into nineteen orders; as the Coleoptera, containing the beetles; the Lepidoptera, containing the butterflies and moths; the Hymenoptera containing the bees, ants and wasps, etc. Four or five of these orders will be of more or less interest to us.

The order Diptera, or two-winged flies, is the most important because to this belong the mosquitoes which transmit malaria and yellow fever, and the house-fly that has come into prominence since it has been found to be such an important factor in the distribution of typhoid and other diseases.

FLIES

The order Diptera is divided into sixty or more families, many of which contain species of considerable economic importance. For our present consideration the flies may be divided into two groups or sections: those with their mouth-parts fitted for piercing such as the mosquito and horse-fly, and those with sucking mouth-parts such as the house-fly, blow-fly and others.

Some of the species belonging to the first group are among the most troublesome pests not only of man but of our domestic animals as well. Next to the mosquitoes the horse-flies (Fig. 22) are perhaps the best known of these. There are several species known under various names, such as gad-fly, breeze-fly, etc. They are very serious pests of horses and cattle, sometimes also attacking man. Their strong, sharp, piercing stylets enable them to pierce through the toughest skin of animals and through the thin clothing of man. The bite is very severe and irritating, and as the flies sometimes occur in great numbers the annoyance that they cause is often very great indeed. It has often been claimed that these flies as well as the stable-fly and others carry the anthrax bacillus on their proboscis from one animal to another, and although this may not have been definitely proven the evidence is strong enough to make a very good case against the accused. It is interesting to note in this connection that anthrax, a very common disease among the domestic animals and one which may attack man also, was the first disease to be shown to be of bacterial origin. It was only about thirty-five years ago that the investigations of Koch and Pasteur demonstrated that the presence of this particular germ (Bacillus anthracis) was the cause of the disease, and it was early recognized that such biting flies may be important factors in the spread of the disease.

The stable-fly (Fig. 23) (Stomoxys calcitrans) which looks very much like the house-fly and, as will be noted later, frequently enters houses, is often an important pest of horses and cattle. Its blood-sucking habit makes it quite possible that it too may be concerned in carrying anthrax and other diseases.

In a later chapter it will be shown how the tsetse-fly, which is somewhat like the stable-fly, is responsible for the spread of the disease known as the sleeping sickness. This disease is caused by a Protozoan parasite, a trypanosome, which is transmitted from one host to another by the tsetse-fly.

In Southern Asia and in parts of Africa there is a very serious disease of horses known as surra which is caused by a similar parasite (Trypanosoma evansi). This parasite attacks horses, mules, camels, elephants, buffaloes and dogs, and has been recently imported into the Philippines. It is supposed that flies belonging to the same genus as the horse-fly (Tabanus and others), and the stable-fly (Stomoxys) and the horn-fly (H 鎚 atobia) are responsible for the spread of the disease.

Nagana is one of the most serious diseases of domestic animals in Central and Southern Africa. In some sections it is almost impossible to keep any kind of imported animals on account of this disease which is caused by a parasite (Trypanosoma brucei) similar to the one causing surra. This parasite is to be found in several different kinds of native animals which seem to be practically immune but are always a source of danger when other animals are introduced. Two or three species of tsetse-flies are responsible for the transmission of this disease.

Another group of flies much smaller but more numerous and much more insistent are the black-flies or buffalo-gnats (Fig. 24). For more than a century these little flies have been recognized as among the most serious pests of stock, particularly in the south where, besides the actual loss by death of many animals yearly, the annoyance is so great as to sometimes make it impossible to work in the field. Human beings are often attacked, and as the bite is poisonous and very painful great suffering may result and cases of deaths from such bites have been reported.

Belonging to another family, and smaller, but much like the buffalo-gnat in habits, are the minute little "punkies" or "no-see-ums" which sometimes occur in great swarms in certain regions where they make life a burden to man and beast. While it has not been shown that either the buffalo-gnats or the punkies are responsible for the transmission of any disease, their habits of feeding on so many different kinds of wild and domestic animals as well as on man makes it possible for them to act as carriers of parasites that might under proper conditions become of serious importance. Then, too, the irritation caused by the bites of these insects usually causes scratching which may result in abrasions of the skin that open the way for various harmful germs, particularly those causing skin diseases.

Coming now to the group containing the house-flies and related forms we find a number that are of interest on account of the suffering that they may cause, particularly in their larval stages.

The screw-worm flies (Chrysomyia macellaria) are among the most common and important of these (Fig. 25). These "gray flies," as they are sometimes called, lay a mass of three or four hundred eggs on the surface of wounds.

The larvaewhich in a few hours hatch from these make their way directly into the wound where they feed on the surrounding tissue until full grown when they wriggle out and drop to the ground where they transform to the pupa and later to the adult fly. Of course their presence in the wounds is very distressing to the infected animal, and great suffering results. Slight scratches that might otherwise quickly heal often become serious sores because of the presence of these larvae

Many cases are recorded of these flies laying their eggs in the ears or nose of children or of persons sleeping out of doors during the day. Especially is this apt to occur if there are offensive discharges which attract the fly. In such cases the larvaeburrow into the surrounding tissues, devouring the mucous membranes, the muscles and even the bones, causing terrible suffering and usually, death. The larvaein such situations may be killed with chloroform and, if the case is attended to before they have destroyed too much of the tissues, recovery usually occurs.

The blow-flies (Fig. 26) (Calliphora vomitoria) and the blue-bottle flies (Fig. 27), (Lucilia spp.) and the flesh-flies (Fig. 28) (Sarcophaga spp.) all have habits somewhat like the screw-worm fly. Any of them may lay their eggs in wounds on man or animals with the same serious results.

The flesh-fly instead of laying eggs deposits the living larvaeupon meat wherever it is accessible, and as these develop with astonishing rapidity they are able to consume large quantities of flesh in a remarkably short time. In this way they may be of some importance as scavengers, but it is better to get rid of the waste in other ways than to leave it for a breeding-place for flies that are capable of causing so much damage and suffering.

Not infrequently the larvaeof certain flies are to be found in the alimentary canal where as a rule they do no particular damage. Altogether the larvaeof over twenty different species of flies have been found in or expelled from the human intestinal canal. In Europe, the majority of these larvaebelong to a fly which looks very much like the house-fly except that it is somewhat smaller and so is often known as "the little house-fly" (Fig. 29) (Homalomyia canicularis). The same species is very common in the United States, frequently occurring in houses. Under certain conditions it may even be more abundant than the house-fly. It is believed that the larvaein the intestinal canal come from eggs that have been deposited on the victim while using an outdoor privy where the flies are often very abundant. Instances are also on record where these larvaehave been discharged from the urethra.

Another fly (Ochromyia anthropophaga) occurring in the Congo region has a blood-sucking larvaewhich is known as the Congo floor-maggot. The fly which is itself not troublesome deposits its eggs in the cracks and crevices of the mud floors of the huts. The larvaewhich hatch from these crawl out at night and suck the blood of the victim that may be sleeping on the floor or on a low bed.

BOT-FLIES

Another group of flies known as the bot-flies (Fig. 30) have their mouth-parts rudimentary or entirely wanting so of course they themselves cannot bite or pierce an animal. Nevertheless they are the source of an endless amount of trouble to stockmen and sometimes even attack man. Although these flies cannot bite, the presence of even a single individual may be enough to annoy a horse almost to the end of endurance. Horses seem to have an instinctive fear of them and will do all in their power to get rid of the annoying pests.

The eggs of the house bot-fly are laid on the hair of the legs or some other part of the body. The horse licks them off and they hatch and develop in the alimentary canal of their host. Sometimes the walls of the stomach may be almost covered with them thus of course seriously interfering with the functions of this organ. When full grown the larvaepass from the host and complete their transformation in the ground.

The bot-flies of cattle or the oxwarbles (Fig. 31) gain an entrance into the alimentary canal in the same way, that is, by the eggs being licked from the hairs on the body where they have been laid by the adult fly. But instead of passing on into the stomach they collect in the esophagus and later make their way through the walls of this organ and through the tissues of the body until they at last reach a place along the back just under the skin. Here as they are completing their development they make more or less serious sores on the backs of the infested animals. The hides on such animals are rendered nearly valueless by the holes made by the larvae When fully mature they drop to the ground and complete their transformations.

The sheep bot-flies (Fig. 32) lay their eggs in the nostrils of sheep. The larvaepass up into the frontal sinuses where they feed on the mucus, causing great suffering and loss. Many other species of animals are infested with their own particular species of bots. Several instances are recorded where the oxwarble has occurred in man, always causing much suffering and sometimes death.

One or more species of bot-flies occurring in the tropical parts of America frequently attack man. The early larval stage soon after it has entered the skin is known as the Ver macaque. Later stages as torcel or Berne. The presence of the larvaeproduces very painful and troublesome sores. It is supposed that the adult flies (one species of which is Dermatobia cyaniventris) lay their eggs on the skin which the larvaepenetrate as soon as they hatch. It has also been suggested that they might reach the subcutaneous tissue by migrating from the alimentary canal as do some of the other bot-flies. A very serious eye disease, Egyptian opthalmia, is known to be spread by the house-flies and others. These flies are often abundant about the eyes, especially of children suffering from this disease. It is suspected that certain small flies (Oscinid? in the southern part of the United States are responsible for the spread of disease known as "sore eye."

FLEAS

The fleas used to be considered as degenerate Diptera and were placed with that group but they are now classed as a separate order (Siphonaptera). Within recent years these little pests have come into special prominence on account of their importance in connection with the spread of the plague. The fact that they are so abundant everywhere and that they will so readily pass from one host to another makes the possibility of their spreading infectious diseases very great. Besides the kinds that are concerned in the transmission of plague, which are discussed in another chapter, there are many other kinds infesting various wild and domesticated animals and a few attacking birds.

One of the most important of these is the jigger-flea or chigoe (Dermatophilus penetrans, Fig. 33). Various other names such as chigger-flea, sand-flea, jigger, chigger are also applied to this insect as well as to a minute red mite that burrows into the skin in much the same way as the female of the flea. So although they are entirely different creatures you can never tell from the common name, whether it is the flea or the mite that is being referred to. Both the male and female jigger-fleas feed on the host and hop on or off as do other fleas, but when the female is ready to lay eggs (Fig. 34), she burrows into the skin. Her presence there causes a swelling and usually an ulcer which often becomes very serious, especially if the insect should be crushed and the contents of the body escape into the surrounding tissue.

These little pests are found throughout tropical and subtropical America and have been introduced into Africa and from there have spread to India and elsewhere. They attack almost all kinds of animals as well as many birds, being of course a source of great annoyance and no inconsiderable loss. They are more apt to attack the feet of men, especially those who go barefooted. Sometimes they occur in such numbers as to make great masses of sores.

On account of being such general feeders they are difficult to control, but some relief may be obtained by keeping the houses and barns as free as possible from dirt and rubbish and by sprinkling the breeding-places of the pest with pyrethrum powder or carbolic water. Those that gain an entrance into the skin should be cut out, care being taken to remove the insect entire.

BEDBUGS

In the order Hemiptera, or the true "bugs" in an entomological sense, we find a few forms that may carry disease. The bedbug (Fig. 35) (Cimex lectularis) has been accused of transmitting plague, relapsing fever and other diseases. Very recent investigations show that the common bedbug of India (Cimex rotundatus) harbors the parasite that causes the disease known as kala azar, and there is no doubt that it transmits the disease.

LICE

The sucking lice (Fig. 36) which also belong to this order are suspected of carrying some of these same diseases. It is thought that the common louse on rats (Hatopinus spinulosus) is responsible for the spread from rat to rat of a certain parasite. (Trypanosoma lewisi), which, however, does not produce any disease in the rats, but if they are capable of acting as alternative hosts for such parasites, it is quite possible that they may also carry disease-producing forms.

HOW INSECTS MAY CARRY DISEASE GERMS

Insects may carry the germs or parasites which cause disease in a purely mechanical or accidental way, that is, the insect may in the course of its wanderings or its feeding get some of the germs on or in its body and may by chance carry these to the food, or water, or directly to some person who may become infected. Thus the house-fly may carry the typhoid germs on its feet or in its body and distribute them in places where they may enter the human body.

Several other flies as well as fleas, bedbugs, ticks, etc., may also carry disease germs in this mechanical way. While this method of transmission is just as dangerous as any other, and possibly more dangerous because more common, another method in which the insect is much more intimately concerned is more interesting from a biological standpoint at least and will be discussed more fully in the chapters on malaria, yellow fever and elephantiasis.

In these cases the insect is one of the necessary hosts of the parasite, which cannot go on with its development or pass from one patient to another unless it first enters the insect at a certain stage of its life-history.

BABY-BYE.

1. Baby-Bye, Here's a fly; We will watch him, you and I. How he crawls Up the walls, Yet he never falls! I believe with six such legs You and I could walk on eggs. There he goes On his toes, Tickling Baby's nose.

CHAPTER V

HOUSE-FLIES OR TYPHOID-FLIES

The page shown in Fig. 37 was copied from one of our old second readers and shows something of the spirit in which we used to regard the house-fly. A few of them were nice things to have around to make things seem "homelike." Of course they sometimes became too friendly during the early morning hours when we were trying to take just one more little nap or they were sometimes too insistent for their portion of the dinner after it had been placed on the table, but a screen over the bed would help us out a little in the morning and a long fly-brush cut from a tree in the yard or made of strips of paper tacked to a stick or, still more fancy, made of long peacock plumes, would help to drive them from the table. Those that were knocked into the coffee or the cream could be fished out; those that went into the soup or the hash were never missed!

Not only were the flies regarded as splendid things with which to amuse the baby, but they were thought to be very useful as scavengers as they were often seen feeding on all kinds of refuse in the yard. Then, too, they seemed to be cleanly little things, for almost any time some of them could be seen brushing their heads and bodies with their legs and evidently having a good clean-up. More than that it never occurred to us that it would be possible to get rid of them even should it be thought advisable, for they came from "out doors," and who could kill all the flies "out doors"?

Fortunately, or otherwise, these halcyon days have gone by and the common, innocent, friendly little house-fly is now an outcast convicted of many crimes and accused of a long list of others (Fig. 38).

Its former friends have become its sworn enemies. The foremost entomologist of the land has suggested that we even change its name and give it one that would be more suggestive of the abhorence with which we now look upon it.

And all these changes have come about because science has turned the microscope on the house-fly and men have studied its habits. We know now that as the fly is "tickling baby's nose" it may be spreading there where they may be inhaled or where they may be taken into the baby's mouth thousands of germs some of which may cause some serious disease. We know that as they are buzzing about our faces while we are trying to sleep they may, unwittingly, be in the same nefarious business, and we know that as they sip from our cups with us or bathe in our coffee or our soup or walk daintily over our beefsteak or frosted cake they are leaving behind a trail of filth and bacteria, and we know that some of these germs may be and often are the cause of some of our common diseases. As the typhoid germs are very often distributed in this way, Dr. Howard has suggested that the house-fly shall be known in the future as the typhoid-fly, not because it is solely responsible for the spread of typhoid, but because it is such an important factor in it and is so dangerous from every point of view. The names "manure fly" and "privy fly" have also been suggested and would perhaps serve just as well, as the only object in giving it another name would be to find a more repulsive one to remind us constantly of the filthy and dangerous habits of the fly.

STRUCTURE

In order that we may better understand why it is that the house-fly is capable of so much mischief, let us consider briefly a few points in regard to its structure, its methods of feeding and its life-history.

The large compound eyes are the most conspicuous part of the head (Fig. 39). In front, between the eyes, are the three-jointed antennae the last joint bearing a short, feathery bristle. From the under side of the head arises the long, fleshy proboscis (Fig. 40). When this is fully extended it is somewhat longer than the head; when not distended and in use it is doubled back in the cavity on the under side of the head. About half-way between the base and the middle is a pair of unjointed mouth-feelers (maxillary palpi). At the tip are two membranous lobes (Fig. 41) closely united along their middle line. These are covered with many fine corrugated ridges, which under the microscope look like fine spirals and are known as pseudotrachea Thus it will be seen that the house-fly's mouth-parts are fitted for sucking and not for biting. Its food must be in a liquid or semi-liquid state before it can be sucked through the tube leading from the lobes at the tip up through the proboscis and on into the stomach. If the fly wishes to feed on any substance such as sugar, that is not liquid, it first pours out some saliva on it and then begins to rasp it with the rough terminal lobes of the proboscis, thus reducing the food to a consistency that will enable the fly to suck it up. Many people think that house-flies can bite and will tell you that they have been bitten by them. But a careful examination of the offender, in such instances, will show that it was not a house-fly but probably a stable-fly, which does have mouth-parts fitted for piercing.

The thorax bears the two rather broad, membranous wings (Fig. 42) which have characteristic venation. Three of these veins end rather close together just before the tip of the wing, the posterior one of the group being bent forward rather sharply a short distance from the tip. The stable-fly has this vein slightly curved forward but not nearly so conspicuously (Fig. 43).

Nearly all the other flies that are apt to be mistaken for the house-fly do not have this vein curved forward. The wings, although apparently bare, are covered with a fine microscopic pubescence. Among these fine hairs on the wing as well as among similar fine ones and coarser ones all over the body, particles of dust and dirt or filth (Fig. 44) or, what interests us more just now, thousands of germs may find a temporary lodgment and later be scattered through the air as the insect flies. Or they may get on our food as the fly feeds or while it rests and combs its body with the rows of coarse hairs on its legs.

The legs are rather thickly covered with coarse hairs or bristles and with a mat of fine, short hairs. On some of the segments the larger hairs are arranged in rows and are used as a sort of comb with which the fly combs the dirt from the rest of its body. The last segment (Fig. 45) of the leg bears at its tip a pair of large curved claws and a pair of membranous pads known as the pulvill? On the under side of the pulvill?are innumerable minute secreting hairs (Fig. 46) by means of which the fly is able to walk on the wall or ceiling or in any position on highly-polished surfaces.

HOW THEY CARRY BACTERIA

These same little pads, with their covering of secreting hairs, are perhaps the most dangerous part of the insect for they cannot help but carry much of the filth over or through which the fly walks, and as this may be well stocked with germs the danger is at once apparent.

As the result of a series of carefully planned experiments it has been demonstrated that the number of bacteria on a single fly may range all the way from 550 to 6,600,000 with an average for the lot experimented with of about one and one-fourth million bacteria to each fly. Now where do all these bacteria come from? Necessarily from the place where the fly breeds or where it feeds.

LIFE-HISTORY AND HABITS

The eggs of the house-fly may be laid on almost any kind of decaying or fermenting material. If this is kept moist and a proper temperature maintained the larvaeor maggots (Fig. 47) that hatch from the eggs may develop. As a rule, however, these requirements are found only under certain conditions and are ordinarily found only in manure heaps or in privy vaults or latrines. All observers agree that the female fly prefers to deposit her eggs in horse manure when this can be found and when this is piled in heaps in the barn-yard (Fig. 48) or in the field the heat caused by the decay and fermentation makes ideal conditions for the development of the larvae Cow manure may serve as a breeding-place to a limited extent. The flies are immediately attracted to human excrement and breed freely in it when opportunity offers. Decaying vegetables or fruit, fermenting kitchen refuse and other materials sometimes also serve as breeding-places.

In suitable places in warm weather the eggs will hatch in from eight to twelve hours and the larvaewill become fully developed in from eight to fourteen days. They then change to pupae(Fig. 50) in which stage they may remain for another eight to twenty days when the adult flies will emerge. These figures must necessarily be indefinite because the weather and other conditions always vary. Under the most favorable conditions of moisture and temperature it is probably never less than eight days from egg to adult fly and under unfavorable conditions it may be as long as six weeks.

The larvaethrive best when the manure is kept quite wet. I have often found them in almost incredible numbers in stables that had not been cleaned for some time. The horses standing there at night added fresh material and kept it just wet enough to make conditions almost ideal (Fig. 49).

The pupaeare usually found where the manure is a little dryer, but it must not be too dry. When the flies issue from the pupaethey push their way up to the surface where they remain for a short time and allow the body to harden and the wings to dry before they fly away to other manure or, as too often happens, to some near-by kitchen or restaurant or market place.

Of course it is impossible for them to issue from this filth without more or less of it clinging to their bodies. Now if these flies would breed only in barn-yard manure and fly directly from the stable to the house there would be comparatively little reason to complain, at least from a sanitary standpoint, for the amount of barn-yard filth that they carried to our food would be of little consequence. But when they breed in privy vaults or similar places, or visit such places before coming into the house or dairy or market place the results may be much more serious.

FLIES AND TYPHOID

It has been abundantly demonstrated that the excrement or the urine of a typhoid patient may contain virulent germs for some time before he is aware that he has the disease, and it has been shown that the germs may be present for weeks or months, and in some cases even years after the patient has recovered. If a fly breeds in such infected material, or feeds or walks on it, it is very apt to get some of the germs on its body where they may retain their virulence for some time, and should it visit our food while covered with these germs some of them would probably be left there where they might produce serious results. More than that. If the fly should feed on such infected material the typhoid germs would go on developing in the intestine of the fly and would be passed out with the feces in which they retain their virulence for some days. In other words, the too familiar "fly-specks" are not only disgusting, but may be a very grave source of danger. It will be seen that in this way several members of a community might become infected with the typhoid germs before anyone was aware that there was a case of typhoid or a "bacillus carrier" in the neighborhood.

One more example out of the scores that might be cited to show how the fly may carry typhoid germs. They may enter the sick chamber in the home or in the hospital and there gain access to the typhoid germs. These they may carry to other parts of the house or to near-by houses, or the flies may light on passing carriages or cars and be carried perhaps for miles before they enter another house and contaminate the food there.

These are hypothetical cases, but they illustrate what is taking place hundreds of times every season all over the world wherever typhoid fever and flies occur, and no country or race is known to be immune from typhoid, and the fly is found "wherever man is found."

In the summer of 1898 a commission was appointed to investigate the prevalence of typhoid fever in the United States Army Concentration Camps. The following are some of the conclusions as reported by Dr. Vaughan:

"FLIES UNDOUBTEDLY SERVED AS CARRIERS OF THE INFECTION

"My reasons for believing that flies were active in the dissemination of typhoid may be stated as follows:

"a. Flies swarmed over infected fecal matter in the pits and then visited and fed upon the food prepared for the soldiers at the mess tents. In some instances where lime had recently been sprinkled over the contents of the pits, flies with their feet whitened with lime were seen walking over the food.

"b. Officers whose mess tents were protected by means of screens suffered proportionately less from typhoid fever than did those whose tents were not so protected.

"c. Typhoid fever gradually disappeared in the fall of 1898, with the approach of cold weather, and the consequent disabling of the fly.

"It is possible for the fly to carry the typhoid bacillus in two ways. In the first place, fecal matter containing the typhoid germ may adhere to the fly and be mechanically transported. In the second place, it is possible that the typhoid bacillus may be carried in the digestive organs of the fly and may be deposited with its excrement."

In Dr. Daniel D. Jackson's report to the Merchants' Association of New York on the "Pollution of New York Harbor as a Menace to the Health by the Dissemination of Intestinal Diseases Through the Agency of the Common House-fly," he shows graphically that the prevalence of typhoid and other intestinal diseases is coincident with the prevalence of flies, and that the greatest number of deaths from such diseases occurs near the river front where the open or poorly constructed sewers scatter the filth where the flies can feed on it, or along the wharves with their inadequate accommodations and the resulting accumulation of filth.

FLIES AND OTHER DISEASES

Not only is the house-fly an important factor in the dissemination of typhoid fever, but it has been definitely shown that it is capable of transmitting several other serious diseases.

The evidence that flies carry and spread the deadly germs of cholera is most conclusive. The germs may be carried on the body where they will live but a short time, or they may be carried in the alimentary canal where they will live for a much longer period and are finally deposited in the fly-specks where they retain their virulence for some time. Flies that had been allowed to contaminate themselves with cholera germs were allowed access to milk and meat. In both cases hundreds of colonies of the germs could later be recovered from the food. As with the typhoid germs milk seems to be a particularly good medium for the development of the cholera germs. In several of the experiments that have been made along this line the milk has been readily infected by the flies visiting it.

Of course an outbreak of cholera is of rare occurrence in our country, but unfortunately this is not so in regard to some other intestinal diseases such as diarrhea and enteritis which annually cause the death of many children, especially bottle-fed babies. Those who have made close studies of the way in which these diseases are disseminated are convinced that the flies are one of the most important factors in their spread.

It has long been observed that flies are particularly fond of sputum and will feed on it on the sidewalk, in the gutter, the cuspidor or wherever opportunity offers. It is well known, too, that the sputum of a consumptive contains myriads of virulent tubercular germs. A fly feeding and crawling over such material must necessarily get some of it on its body, and as it dries and the insect flies about the germs will be distributed through the air, possibly over our food. It has been shown that the excretion from a fly that has fed on tubercular sputum contains tubercular bacilli that may remain virulent for at least fifteen days. Thus we see again the danger that may lurk in the too familiar "fly-specks."

Although it is generally supposed that the flea is solely responsible for the spread of the bubonic plague and no doubt is the principal distributing agent, the fact must not be overlooked that the house-fly may also be of considerable importance in this connection. Carefully planned experiments have shown that flies that have become infected by being fed on plague-infected material may carry the germs for several days and that they may die of the disease. During plague epidemics flies may become infected by visiting the sores on human or rat victims or by feeding on dead rats or on the excreta of sick patients, and an infected fly is always a menace should it visit our food or open wounds or sores. Anthrax bacilli are carried about and deposited by flies showing the possibility of the disease being spread in this way.

Some believe that leprosy, smallpox and many other diseases are carried by the house-fly, so it is little wonder that it is fast losing its standing as a household companion and that we are beginning to regard it not only as a nuisance but as a source of danger which should no longer be tolerated in any community.

Of course only a small per cent of the flies that visit our food in the dairies or market places or kitchens actually carry dangerous diseases, but they are all bred in filth and it is not possible without careful experiments or laboratory analysis to determine whether any of the germs among the millions that are on their bodies are dangerous or not. The chances that they may be are too great. The only safe way is to banish them all or to see that all of our food is protected from them.

FIGHTING FLIES

Screens and sticky fly-paper have their places and give some little relief in a well-kept house. But of what use is it to protect your food after it has entered your home if in the stores, in the market place, in the dairy barn, or dairy wagon, in the grocers' and butchers' cart, it has been exposed to contamination by hundreds of flies that have visited it.

The problem is a larger one than keeping the house free from flies; larger but not more difficult, for the remedy is simple, effective, practicable and inexpensive. Destroy their breeding-places and you will have no flies. As the flies breed principally in manure the first remedial measure is to see that all manure is removed from the barn-yard at least once a week and spread over the fields to dry, for the flies cannot breed in the dry manure. If it is not practicable to remove it this often the manure should be kept in a bin that is closed so tight that no flies can get into it to lay their eggs. Sometimes the manure may be treated with some substance such as kerosene, crude oil, chlorid of lime, tobacco water or mixture of two or more of these and thus rendered unsuitable for the flies to breed in, but in general practice none of them has been found very satisfactory for the treatment is either not thorough enough or is too expensive of time and material.

Outdoor privies and cesspools must be carefully attended to. The latter can be easily covered so no flies can get in and if the filthy and in every way dangerous pit under the privy be filled and the dry-earth closet substituted one of the greatest sources of danger, especially in the country and in towns with inadequate sewerage facilities, will be done away with. After these things are done there remain only the garbage cans and the rubbish heaps to look after.

Of course your neighbor must keep his place clean too, for his flies are just as apt to come into your house as his, so the problem becomes one for the whole community.

Almost all cities and many of the smaller towns have ordinances which if enforced would afford adequate protection from flies, but they are seldom if ever rigidly enforced and it yet remains for some enterprising town to be able to advertise itself as a "speckless town" as well as a "spotless town."

AN EXPERT'S OPINION

In a recent important bulletin issued by the Bureau of Entomology, Dr. L.O. Howard discusses the economic importance of several of the insects that carry disease. I wish to quote two or three paragraphs from the pages in which he discusses the house-fly or typhoid fly to show the opinion of this excellent authority in regard to this pest.

"Even if the typhoid or house fly were a creature difficult to destroy, the general failure on the part of communities to make any efforts whatever to reduce its numbers could properly be termed criminal neglect; but since, as will be shown, it is comparatively an easy matter to do away with the plague of flies, this neglect becomes an evidence of ignorance or of a carelessness in regard to disease-producing filth which to the informed mind constitutes a serious blot on civilized methods of life."

On another page:

"We have thus shown that the typhoid or house fly is a general and common carrier of pathogenic bacteria. It may carry typhoid fever, Asiatic cholera, dysentery, cholera morbus, and other intestinal diseases; it may carry the bacilli of tuberculosis and certain eye diseases. It is the duty of every individual to guard so far as possible against the occurrence of flies upon his premises. It is the duty of every community, through its board of health, to spend money in the warfare against this enemy of mankind. This duty is as pronounced as though the community were attacked by bands of ravenous wolves."

Again:

"A leading editorial in an afternoon paper of the city of Washington, of October 20, 1908, bears the heading, 'Typhoid a National Scourge,' arguing that it is to-day as great a scourge as tuberculosis. The editorial writer might equally well have used the heading 'Typhoid a National Reproach,' or perhaps even 'Typhoid a National Crime,' since it is an absolutely preventable disease. And as for the typhoid fly, that a creature born in indescribable filth and absolutely swarming with disease germs should practically be invited to multiply unchecked, even in great centers of population, is surely nothing less than criminal."

The whole bulletin (No. 78, Bureau of Entomology) should be read and studied by all who are interested in this subject.

OTHER FLIES

Occasionally other flies looking more or less like the house-fly are seen in houses. Some of these have the same type of sucking mouth-parts and have habits very similar to the house-fly, but as they are usually much less common and as nearly all that has been said in regard to the house-fly would apply equally well to them and as the same measures should be adopted in fighting them they need not be discussed further here.

I have already called attention to the fact that a fly which looks very much like the house-fly is sometimes found in the house and will often bite severely. It has quite a different style of beak, one that is fitted for piercing so it may suck the blood of its victim (Fig. 51). As these flies often seem to be more persistent before a rain the weather prophet will tell you that "It is surely going to rain for the house-flies are beginning to bite."

These stable-flies, as they are called, are great pests of cattle and horses in some sections. It is thought that they are important factors in the spread of some of the diseases of domestic animals, and their habit of sometimes attacking human beings makes it possible for them to carry certain disease germs from animals to man or from man to man.

CHAPTER VI

MOSQUITOES

Mosquitoes are no more abundant now than they have been in the past, but when Linn 鴶 s in 1758 made his list of all the animals known to exist at that time he catalogued only six species of mosquitoes. Only a few years ago, 1901, Dr. Theobald of the British Museum published a book on the mosquitoes of the world in which he listed three hundred and forty-three kinds. Soon other volumes appeared, adding more species, and systematists everywhere have been describing new ones until now the total number of described species is probably over five hundred, more than sixty of which occur in the United States.

This shows only one phase of the great interest that has been taken in the mosquitoes since the discovery of their importance as carriers of disease. Not only have they been studied from a systematic standpoint but an endless amount of work has been done and is being done in studying their development, habits, and structure until now, if one could gather together all that has been written about mosquitoes in the last ten or twelve years he would have a considerable library.

Those who are particularly interested in the group will find some of these books and papers easily accessible, so there may be given here only a brief summary of the more important facts in regard to the structure and habits of the mosquitoes in order that we may more readily understand the part that they play in the transmission of diseases and see the reasonableness of the recommendations in regard to fighting them.

THE EGGS

Mosquito eggs are laid in water or in places where water is apt to accumulate, otherwise they will not hatch. Some species lay their eggs in little masses (Fig. 52) that float on the surface of the water, looking like small particles of soot. Others lay their eggs singly, some floating about on the surface, others sinking to the bottom where they remain until the young issue. Some of the eggs may remain over winter, but usually those laid in the summer hatch in thirty-six to forty-eight hours or longer according to the temperature.

THE LARVAE

When the larvaeare ready to issue they burst open the lower end of the eggs and the young wrigglers escape into the water. The larvaeare fitted for aquatic life only, so mosquitoes cannot breed in moist or damp places unless there is at least a small amount of standing water there. A very little will do, but there must be enough to cover the larvaeor they perish.

The head of the larvaeof most species is wide and flattened. The eyes are situated at the sides, and just in front of them is a pair of short antennaewhich vary with the different species.

The mouth-parts too vary greatly according to the feeding habits. Some mosquito larvaeare predaceous, feeding on the young of other species or on other insects. These of course have their mouth-parts fitted for seizing and holding their prey. Most of the wrigglers, however, feed on alg? diatoms, Protozoa and other minute plant or animal forms which are swept into the mouth by curious little brush-like organs whose movements keep a stream of water flowing toward the mouth.

Another group containing the Anopheles are intermediate between these two and have mouth-parts fitted for feeding on minute organisms as well as for attacking and holding other larger things.

A few kinds feed habitually some distance below the surface, others on the bottom, while still others feed always at the surface. With one or two exceptions, the larvaemust all come to the surface to breathe (Figs. 53-57). Most species have on the eighth abdominal segment a rather long breathing-tube the tip of which is thrust just above the surface of the water when they come up for air. In this tube are two large vessels or tracheawhich open just below the tip of the tube and extend forward through the whole length of the body, giving off branches here and there that divide into still smaller branches until every part of the body is reached by some of the small divisions of this tracheal system that carries the oxygen to all the tissues. The length of the breathing-tube is correlated with the feeding-habits of the larvae Anopheles larvaewhich feed at the surface have very short tubes (Fig. 58), others that feed just below the surface have breathing-tubes as long or very much longer than the ninth abdominal segment. The last segment has at its tip four thin flat plates, the tracheal gills. These too are larger or smaller according to the habits of the larvae Those species that feed close to the surface and have the tip of the breathing-tube above the surface most of the time have very small tracheal gills, while those that feed mostly on the bottom have them well developed.

When first hatched the larvaeare of course very small. If the weather is warm and the food is abundant they grow very rapidly. In a few days the outer skin becomes rather firm and inelastic so it will not allow further growth. Then a new skin forms underneath and the old skin is cast off. This process of casting off the old skin is called molting, and is repeated four times during the one, two, three or more weeks of larval life.

PUPA

With the fourth molt the active feeding larva changes to the still active but non-feeding pupa (Fig. 59). The head and thorax are closely united and a close inspection will reveal the head, antennae wings and legs of the adult mosquito folded away beneath the pupal skin. Instead of the breathing-tube on the eighth segment of the abdomen as in the larva, the pupa has two trumpet-shaped tubes on the back of the thorax through which it now gets its air from above the surface. The pupal stage lasts from two to five or six days or more. When the adult is ready to issue the pupal skin splits along the back and the mosquito gradually and slowly issues. It usually takes several minutes for the adult to issue and for its wings to become hard enough so it can fly. In the meantime, it is resting on the old pupal skin or on the surface of the water, where it is entirely at the mercy of any of its enemies that might happen along and is in constant danger of being tumbled over should the water not be perfectly smooth.

THE ADULT

The adult mosquito is altogether too familiar an object to need description, but it is necessary that we keep in mind certain particular points in regard to its structure, in order that we may better understand how it is that it is capable of transmitting disease.

If we examine closely the antennaeof a number of mosquitoes that are bothering us with their too constant attentions we shall see that they all look very much alike (Fig. 62), small cylindrical joints bearing whorls of short fine hairs. But if we examine a number of mosquitoes that have been bred from a jar or aquarium we will find two types of antennae the one described above belonging to the female. The antennaeof the male (Fig. 63) are much more conspicuous on account of the whorl of dense, fine, long hairs on each segment. Another interesting difference in the antennaeis to be noted in the size of the first joint. In both sexes it is short and cup-shaped, but in the male it is somewhat larger. This basal segment contains a highly complex auditory organ which responds to the vibrations of the whorls of hairs on the other segments. Interesting experiments have shown that these hairs vibrate best to the pitch corresponding to middle C on the piano, the same pitch in which the female "sings." Of course mosquitoes and other insects have no voice as we ordinarily understand the word, but produce sound by the rapid vibration of the wings or by the passage of air through the openings of the trachea The males and females are thus easily distinguished and, as we shall see later, this is of some importance for only the females can bite. The males and females differ in another way. Just below the antennaeand at the sides of the proboscis or beak is a pair of three-to five-jointed appendages, the maxillary palpi or mouth-feelers which in the females of most species are very short (Fig. 64) while in the males they are usually as long as the proboscis (Fig. 65). The females of Anopheles and related forms have palpi quite as long as the males, but they are slender throughout while the male palpi are usually somewhat enlarged toward the tip and bear more or less conspicuous patches of rather long hairs or scales.

THE MOUTH-PARTS

The mouth-parts of the mosquito are of course of particular interest to us. At first they appear to consist of a long slender beak or proboscis, but by dissecting and examining with a microscope we find this beak to be made up of several parts (Fig. 66). The labium, which is the largest and most conspicuous, is apparently cylindrical but is grooved above throughout its length. At the tip of the labium are the labell? two little lobes which serve to guide the piercing organs. Lying in this groove along the upper side of the labium are six very fine, sharp-pointed needles. The uppermost of these, the labrum-epipharynx, or labrum as we will call it, is the largest and is really a hollow tube very slightly open on its under side. Just below this is the hypopharynx, the lateral margins of which are very thin. Down through the median line of the hypopharynx runs a minute duct (Fig. 67, sal) which, though exceedingly small, is of very great importance, for through it is poured the saliva which may carry the malaria germs into the wound made when the mosquito bites. The other four needles consist of a pair of mandibles which are lance-shaped at the tip and a heavier pair of maxill? the tips of which are serrate on one edge.

HOW THE MOSQUITO BITES

When the female mosquito is feeding on man or any other animal the tip of the labium is placed against the surface and the six needles are thrust into the skin, the labell?serving as guides. As they are thrust deeper and deeper the labium is bowed back to allow them to enter. As soon as the wound is made the insect pours out through the tube of the hypopharynx some of the secretion from the salivary glands and then begins to suck up the blood through the hollow labrum into the pharynx and on into the stomach.

The mouth-parts of the male differ in some important respects from those of the female. The hypopharynx is united to the labium, the mandibles are wanting and the maxill?are very much reduced so that the insect is unable to pierce the tough skin of animals. The male feeds on the juices of plants as do the females when they cannot get blood. It is not at all necessary for mosquitoes to have the warm blood of man or other animals. Comparatively few of them ever taste blood. They have been seen feeding on blossoms, ripe fruit, watermelons, plant juices, etc. They are very fond of ripe bananas and are fed on them in the laboratory when we wish to keep mosquitoes for experimental purposes.

THE THORAX

The middle part of the body, called the thorax, is really a strong box with heavy walls for the attachment of the powerful wing and leg muscles. The three pairs of legs are covered with hairs and scales, and their tips are provided with a pair of claws which vary somewhat in the different species. The wings (Fig. 68) are long and narrow with a characteristic venation. Along the veins and the margin of the wings are the scales which readily enable one to distinguish mosquitoes from other insects that may look much like them. In some species these scales are long and narrow, almost hair-like, in others they are quite broad and flat (Fig. 69). Just back of the wings is a pair of balancers, short thread-like processes knobbed at the end. These probably represent the second pair of wings with which most insects are provided, and seem to serve as balancers or orienting organs when the insect is flying. On the sides of the thorax are two small slit-like openings, the breathing-pores. These are the openings into the tracheal or respiratory system.

THE ABDOMEN

The long cylindrical abdomen is composed of eight segments. These are rather strongly chitinized above and below, but a narrow strip along the side is unchitinized. In this strip are situated the abdominal breathing-pores. The tip of the abdomen is furnished with a pair of movable organs, which in the male are variously modified and serve as clasping organs at mating time.

THE DIGESTIVE SYSTEM

The mouth-parts of the mosquito have just been described. It will be remembered that the labrum is provided with a groove. Through this the blood or other food is sucked up by means of a strong-walled pumping organ, the pharynx, situated in the head (Fig. 70). Just back of the pharynx is the esophagus which leads to the beginning of the stomach. Close to its posterior end the esophagus gives off three food reservoirs, two above and a single larger one below. In dissections these will often be seen to be filled with minute bubbles. The stomach reaches from the middle of the thorax to beyond the middle of the abdomen. At its posterior end are given off five long slender processes, the Malpighian tubules which are organs of excretion, acting like the kidneys of higher animals. The hindgut is that portion of the intestine from the stomach to the end of the body.

THE SALIVARY GLANDS

Lying under the alimentary canal in the forward part of the thorax are the salivary glands. There are two sets of these, each having three lobes with a common duct which joins the duct from the other set a short distance before they enter the base of the hypopharynx. Each of these lobes is made up of a layer of secreting cells (Fig. 71) which produces the saliva that is poured into the wound as soon as the insect pierces the skin of the victim, and we shall see, too, that the malarial germs also collect in these glands to be carried by the saliva to the new host.

EFFECTS OF THE BITE

After a mosquito has bitten a person and withdrawn the stylets, a small area about the puncture whitens, then soon becomes pink and begins to swell, then to itch and burn. Some people suffer much more from the bites of mosquitoes than do others. For some such bites mean little or no inconvenience, indeed may pass wholly unnoticed, to others a single bite may mean much annoyance, and several bites may cause much suffering.

After an hour or so the itching usually ceases, but in some cases it continues longer. In some instances little or no irritation is felt until some hours, sometimes as much as a day, after the bite. In such cases the effect of the bite is apt to be severe and to last for several days. Sometimes a more or less serious sore will follow a bite, probably due to infection of the wound by scratching. It is doubtless the saliva that is poured into the wound that causes the irritation. It is frequently asserted that if the mosquito is allowed to drink its fill and withdraw its beak without being disturbed no evil results will follow. Those who hold this theory say that the saliva that is poured into the wound is all withdrawn again with the blood if the mosquito is allowed to feed long enough. There may be some truth in this, but for most of us a bite means a hurt anyway and few will be content to sit perfectly still and watch the little pest gradually fill up on blood.

It is not known just what the action of the saliva is, its composition or reaction on the tissues. It is generally supposed to prevent coagulation of the blood that is to be drawn through the narrow tube of the labrum. Others think that its presence causes a greater flow of blood to the wound. But the sad part of it is, for us at least, that it hurts and may cause malaria and possibly other diseases.

HOW MOSQUITOES BREATHE

Mosquitoes and other insects do not have any nostrils nor do they breathe through any openings on the head. Along the sides of the thorax and abdomen is a series of very minute openings known as the spiracles. Through these the air passes into a system of air-tubes, the trachea There are two main trunks or divisions of the tracheajust inside the body-wall and a number of shorter connecting trunks. From these larger vessels arise a great number of smaller ones which branch and subdivide again and again until all the tissues are supplied by these minute little air-tubes that carry the oxygen to all parts of the body and carry off the waste carbon dioxid. These air-tubes are emptied and filled by the contractions of the walls of the abdomen. When the body-wall contracts the air is forced out of the thin-walled trachea through the spiracles; when the pressure is removed they are refilled by the fresh air rushing in.

THE BLOOD

After a mosquito has been feeding on a man or some other animal it is often so distended that the blood shows rich and red through the thin sides of the walls of the abdomen. This, however, is the blood of the victim and not of the mosquito. The blood of insects is not red but pale yellowish or greenish. It is not confined in definite vessels, but fills all the space inside the body cavity that is not occupied by some of the tissues or organs. It bathes the walls of the alimentary canal and gathers there the nourishment which it carries to all parts of the body. It does not carry oxygen or collect the carbon dioxide as does the blood of higher animals. That work, as we have just seen, is done by the air-tubes. Above the alimentary canal, extending almost the whole length of the abdomen and thorax, is a thin-walled pulsating vessel, the heart. This consists of a series of chambers each communicating with the one in front of it by an opening which is guarded by a valve. When one of these chambers contracts it forces the blood that is in it forward into the next chamber which, in its turn, sends it on. As the walls relax the valves at the sides are opened and the blood that is in the body-cavity rushes in to fill the empty chamber. As these regular rythmical pulsations recur the blood is forced forward through the heart into the head where it bathes the organs there. We shall see in another chapter that the malarial parasite escapes from the walls of the stomach of the mosquito into the blood in the body-cavity and finally reaches the salivary glands. As the heart is constantly driving blood to this part of the body the parasites readily reach the glands from which they finally escape into the new host.

CLASSIFICATION

For our purpose it will not be necessary to try to give a system of classification of all the mosquitoes. Those interested in this phase of the subject will find several books and papers devoted wholly to it. It is quite important, however, that we know something about a few of the more familiar groups and kinds, especially those concerned in the transmission of diseases.

THE ANOPHELES

In pointing out the differences between male and female mosquitoes we noted that in one group, the genus Anopheles, both sexes have long maxillary palpi (Figs. 72, 73). This is the most important character separating this genus from the other common forms and as the Anopheles are the malaria carriers it is important that this difference be remembered. Most of the members of this group have spotted wings (Fig. 74), but as some other common kinds also have spotted wings (Fig. 75) this character will not always be reliable. When an Anopheles mosquito is at rest the head and proboscis are held in one line with the body and the body rests at a considerable angle to the surface on which it is standing. Other kinds rest with the body almost or quite parallel to the surface on which they are standing. So if you find a female mosquito with long mouth-palpi and spotted wings resting at an angle to the surface on which it stands you may be reasonably sure that it is an Anopheles and therefore may be dangerous (Figs. 76, 77, 78, 79).

In the United States there are three species of Anopheles--maculipennis, punctipennis and crucians--which are common in various localities, and one or two other species that so far as known are local or rare.

The Anopheles eggs are not laid in masses as are the eggs of many other mosquitoes, but are deposited singly on the surface of the water where they may be found often floating close together.

The eggs (Figs. 80, 81) are elliptical in outline and are provided with a characteristic membranous expansion near the middle.

The larvaemay be found at the proper season and in the localities where they are abundant in almost any kind of standing water, in clear little pools beside running streams, in the overflow from springs, in swamps and marshy lands, in rain-barrels or any other places or vessels where the water is quiet. They do not breed in brackish water. As they feed largely on the alg?or green scum on the surface of the water they are especially apt to be found where this is present. We have already noted that their positions in the water differ from that assumed by other species (Fig. 82).

As the breathing-tube is very short the larvaemust come close to the surface to breathe, and when they are feeding we find them lying just under and parallel to the surface of the water with their curious round heads turned entirely upside down as they feed on the particles that are floating on the surface (Figs. 83, 84).

The pupaedo not differ very much from the pupaeof other species although the breathing-tubes on the thorax are usually shorter and the creature usually rests with its abdomen closer to the surface, that is, it does not hang down from the surface quite as straight as do other forms (Fig. 85).

The adults may be found out of doors or in houses, barns or other outbuildings. They do not seem to like a draft and consequently will be more apt to frequent rooms or places where there is little circulation of air. Although they are usually supposed to fly and bite only in the evening or at night, they may occasionally bite in the daytime. One hungry female took two short meals from my arm while we were trying to get her to pose for a photograph one warm afternoon.

The female passes the winter in the adult condition, hibernating in any convenient place about old trees or logs, in cracks or crevices in doors or out of doors. In the house they hide in the closets, behind the bureau, behind the head of the bed, or underneath it, or in any place where they are not apt to be disturbed. During a warm spell in the winter or if the room is kept warm they may come out for a meal almost any time.

THE YELLOW FEVER MOSQUITO

Ranking next in importance to Anopheles as a disseminator of disease and in fact solely responsible for a more dreaded scourge, is the species of mosquito now known as Stegomyia calopus. While this species is usually restricted to tropical or semi-tropical regions it sometimes makes its appearance in places farther north, especially in summer time, where it may thrive for a time. The adult mosquito (Fig. 104) is black, conspicuously marked with white. The legs and abdomen are banded with white and on the thorax is a series of white lines which in well-preserved specimens distinctly resembles a lyre. These mosquitoes are essentially domestic insects, for they are very rarely found except in houses or in their immediate vicinity. Once they enter a room they will scarcely leave it except to lay their eggs in a near-by cistern, water-pot, or some other convenient place.

Their habit of biting In the daytime has gained for them the name of "day mosquitoes" to distinguish them from the night feeders. But they will bite at night as well as by day and many other species are not at all adverse to a daylight meal, if the opportunity offers, so this habit is not distinctive. The recognition of these facts has a distinct bearing in the methods adopted to prevent the spread of yellow fever. There are no striking characters or habits in the larval or pupal stages that would enable us to distinguish without careful examination this species from other similar forms with which it might be associated. For some time it was claimed that this species would breed only in clean water, but it has been found that it is not nearly so particular, some even claiming that it prefers foul water. I have seen them breeding in countless thousands in company with Stegomyia scutellaris and Culex fatigans in the sewer drains in Tahiti in the streets of Papeete. As the larvaefeed largely on bacteria one would expect to find them in exactly such places where the bacteria are of course abundant.

The fact that they are able to live in any kind of water and in a very small amount of it well adapts them to their habits of living about dwellings.

So far as known the members of these two genera are the only two that are concerned in the transmission of disease in the United States. In other countries other species are suspected or proven disseminators of certain diseases, but these will be discussed in connection with the particular diseases in later chapters.

OTHER SPECIES

The many other species of mosquitoes that we have may be conveniently divided as to their breeding-habits into the fresh-water and the brackish-water forms. Among the fresh-water kinds some are found principally associated with man and his dwelling places, others live in the woods or other places and so are far less troublesome. Most of these do not fly far. Several of the species that breed in brackish water are great travelers and may fly inland for several miles. Thus the towns situated from one to three or four miles inland from the lower reaches of San Francisco Bay are often annoyed more by the mosquitoes that breed only in the brackish water on the salt marshes than they are by any of the fresh-water forms (Figs. 86, 87). The worst mosquito pest along the coast of the eastern United States and for some distance inland is a species that breeds in the salt marshes.

NATURAL ENEMIES OF MOSQUITOES

In combating noxious insects we learned long ago that often the most efficient, the easiest and cheapest way is to depend on their natural enemies to hold them in check. Under normal or rather natural conditions we find that they are usually kept within reasonable bounds by their natural enemies, but under the artificial conditions brought about by the settling and developing of any district great changes come about. It very often happens that these changes are favorable to the development of the noxious insects and unfavorable to the development of their enemies.

A striking example and one to the point is afforded in the introduction of mosquitoes into Hawaii. Up to 1826 there were no mosquitoes on these islands. It is supposed that they were introduced about that time by some ships that were trading at the islands. Indeed it is claimed that the very ship is known that brought them over from Mexico.

Once introduced they found conditions there very favorable to their development, plenty of standing water and few natural enemies to prey on them, so they increased very rapidly and gradually spread over all the islands of the group. This was the so-called night mosquito, Culex pipiens. Much later another species, Stegomyia calopus, just as annoying and much more dangerous was introduced and has also become very troublesome. We have a few species of top-minnows (Fig. 88) occurring in sluggish streams in the southern part of the United States that are important enemies of the mosquitoes of that region. A few years ago some of these were taken over to Hawaii and liberated in suitable places to see if they would not help solve the mosquito problem there. The fishes seem to be doing well. Already they are destroying many mosquito larvae and there are indications that they are going to do an important work, but of course can be depended on only as an aid.

On account of the various habits of both the larvaeand adults it will never be possible for any natural enemy or group of natural enemies effectively to control the mosquitoes of any region, but as certain of them are important as helpers they deserve to be mentioned.

ENEMIES OF THE ADULTS

Birds devour a few mosquitoes, the night-flying forms being particularly serviceable, but the number thus destroyed is probably so small as to be of little practical importance.

The dragon-flies (Figs. 89, 90, 91) or mosquito hawks have long been known as great enemies of mosquitoes, and they certainly do destroy many of them as they are hawking about places where mosquitoes abound. Dr. J.B. Smith of New Jersey very much doubts their efficiency, but observations made by other scientific men would seem to indicate that they often devour large numbers of mosquitoes during the course of the day and evening.

Spiders and toads destroy a few mosquitoes each night. Certain external and internal parasites destroy a few more, but the sum total of all of these agencies is probably not very considerable, for while the adults may have several natural enemies they are not of sufficient importance to have any appreciable effect on the number of mosquitoes in a badly infested region.

ENEMIES OF THE LARVAEAND PUPAE

The larvaeand pupaeon the other hand have many important enemies. Indeed under favorable conditions these may keep small ponds or lakes quite free from the pests. The predaceous aquatic larvaeof many insects feed freely on wrigglers. The larvaeof the diving beetles which are known as water-tigers are particularly ferocious and will soon destroy all the wrigglers in ponds where they are present (Fig. 92). Dragon-fly larvaealso feed freely on mosquito larvae Whirligig beetles are said to be particularly destructive to Anopheles larvaeand many other insects such as water-boatmen, back-swimmers, etc., feed on the larvaeof various species. A few of these introduced into a breeding-jar with Anopheles larvaewill soon destroy all of them, even the very young bugs attacking larvaemuch larger than themselves.

It is interesting to note that the larvaeof some mosquitoes are themselves predaceous and feed freely on the other wrigglers that may chance to be in the same locality.

Various species of fish are, however, the most important enemies of the mosquitoes. Great schools of tide-water minnows (Fig. 93) are often carried over the low salt-marshes by the extreme high-tides and left in the hundreds of tide pools as the tide recedes. No mosquitoes can breed in a pool thus stocked with these fish. In the fresh-water streams and lakes there are several species of the top-minnows, sticklebacks (Fig. 94), etc., that feed voraciously on mosquito larvae and unless the grass or reeds prevent the fish from getting to all parts of the ponds or lakes very few mosquitoes can breed in places where they are present.

Minute red mites such as attack the house-flies and other insects sometimes attack adult mosquitoes, but they are rarely very abundant. Parasitic roundworms attack certain species. Others suffer more or less from the attacks of various Sporozoan parasites.

FIGHTING MOSQUITOES

 When mosquitoes are bothering us we usually begin by trying to kill the individual pests that are nearest to us. We try to crush them if they bite us; we screen the doors and windows to keep them from the house. In warmer countries the people are a little more hospitable and do not screen the mosquitoes out of the house entirely, but screen the beds for protection at night, and if the mosquitoes get too insistent during the day the bed makes a safe and comfortable retreat. All the mosquitoes in a room may be killed by fumigating with sulphur at the rate of two pounds to the thousand cubic feet of air-space. Pyrethrum is also used largely, but it only stupefies the mosquitoes temporarily instead of killing them. While in that condition they may be swept up and destroyed.

 Various substances are sometimes used as repellants by those who must be in regions where the mosquitoes are abundant. With many of these, however, "the cure is worse than the disease." Smudges are often built to the windward of a house or barn-yard and the smoke from a good smoldering fire will keep a considerable area quite free from mosquitoes. The man who can keep himself enveloped in a cloud of tobacco smoke will not be bothered by mosquitoes. Oil of pennyroyal, oil of tar or a mixture of these with olive oil, and various other concoctions are sometimes smeared over the face and hands. These will furnish protection as long as they last. Dr. Smith says that he has found oil of citronella quite effective and of course less objectionable than the other things usually used. Care should be taken not to get it in the eyes. An ointment made of cedar oil, one ounce; oil of citronella, two ounces; spirits of camphor, two ounces, is said to make a good repellant and is effective for a long time.

FIGHTING THE LARVAE
 All of the efforts directed against the adult mosquitoes are usually of little avail in decreasing the number in any region. It is comparatively easy, however, to fight them successfully in the larval stage. We have seen that standing water is absolutely necessary for mosquitoes to breed in. This makes the problem much simpler than if they could breed in any moist places such

as well-sprinkled lawns, a shady part of the garden, etc. The whole problem of successful campaigns against the mosquitoes resolves itself into the problem of finding and destroying or properly treating their breeding-places. We have seen how certain kinds, such as the yellow fever mosquito, are "domestic" species. They never go far from their breeding-places. If a house is infected by one of these species the immediate premises should be searched for the source. Cisterns, rain-barrels, sewer-traps, cesspools, tubs or buckets of water or old tin cans in out-of-the-way corners, are all suitable places for them to breed in. Cisterns and rain-barrels should be thoroughly screened so that no mosquitoes can get in or out, or the surface should be covered with a film of kerosene which will kill all the larvaein the water when they come to the surface to breathe, and will also kill the females when they come to deposit their eggs. The vent to open cesspools should be thoroughly screened or the surface of the water kept well covered with oil. Water standing in any vessels in the yards should be emptied every week or ten days and the old tin cans destroyed or hauled away. In fighting these domestic species you need be concerned only with your own yard and that of your near-by neighbors. Other species, while also rather local in their distribution, fly much farther than the really domestic ones. In fighting these the region for a considerable distance around must be taken into consideration. Watering-troughs (Fig. 95) that are left filled from week to week, the overflow from such places, and the tracks made in the mud round about them (Fig. 96), small sluggish streams, irrigating ditches, and small ponds or lakes not supplied with fish are excellent breeding-places for several species of mosquitoes including Anopheles and others. The remedy at once suggests itself. The watering-trough can be emptied and renewed every week during the summer time, the overflow can be taken care of in a ditch that will lead it away from the trough to where it will sink into the ground, the banks of the streams or ponds or lakes can be cleared in such a way that fish can get to all parts of the water; most of the small ponds can be drained or their surface may be covered over with a thin film of kerosene. This is best applied as a spray; one ounce to fifteen square feet will suffice. If the oil is simply poured over the surface more will be required.

The fighting of the species that breed on the extensive salt-marshes in many regions is a larger and more difficult problem, but as it is a matter that usually concerns large communities, sometimes whole states, it can be dealt with on a larger scale. The very excellent results that have been accomplished in New

Jersey and on the San Francisco peninsula, and in a smaller way in other places, show what may be done if the community goes about the fight in an intelligent manner. In the fight in New Jersey hundreds of acres of tide-lands have been drained so that they no longer have tide pools standing where the mosquitoes may breed. When it is impracticable to drain them the pools may be sprayed occasionally with kerosene.

The value of the land that is reclaimed by a good system of draining is often enough to pay many times over the cost of draining, thus the mosquitoes are gotten rid of and the land enhanced in value by a single operation.

CHAPTER VII

MOSQUITOES AND MALARIA

Ever since the beginning of history we have records of certain fevers that have been called by different names according to the people that were affected. As we study these names and the various writings concerning the fevers we find that a great group of the most important of them are what we to-day know as malarial fevers. Not only are these ills as old as history but they have been observed over almost the entire inhabited earth. There are certain regions in all countries where malaria does not occur, but almost always it will be found that other regions near by are infected and it very often happens that these infected regions are the most profitable parts of the land, the places where water is plentiful and vegetation is luxuriant. Indeed the coincidence of these two things, low-lying lands with an abundance of water, particularly standing water, and malaria has always been noted and gave rise to the earliest theories in regard to the cause of the disease.

For instance, we find some of the very early writers emphasizing the point that swampy localities should be avoided for they produce animals that give rise to disease, or that the air is poisoned by the breath of the swamp-inhabiting animals.

These views of the origin of the fever prevailed until about the beginning of the eighteenth century when the recently discovered microscope began to reveal the various kinds of animalcul?to be found in decaying material.

In 1718 Lancisi held that the myriads of insects, particularly gnats or mosquitoes, that arose from such swampy regions might carry some of these poisonous substances and by means of their proboscis introduce them into the bodies of the people, and although he had made no experiments to test the assumption he did not consider it impossible that such insects might also introduce the smallest animalcul?into the blood. It took almost two centuries of study and investigation before this guess was proved to be right.

One reason why the mosquitoes were not earlier associated with these diseases was that all who investigated the matter at all turned their attention to the bad condition of the air in these swampy regions. Malaria means bad air. We all know that we can see the mists arising from such regions, particularly in the evening or at night, and as exposure to these mists very often meant an attack of malaria they were naturally supposed to be the cause of the disease. So for a long time the whole attention of investigators was turned toward studying and analyzing these vapors, and various experiments were made which seemed to show conclusively that the malaria was caused only by these emanations. The investigations even went so far that the exact germs that were supposed to cause the fever were separated and experimented with.

THE PARASITE THAT CAUSES MALARIA

The blood had been studied time and again and the characteristic appearance of the blood of a malarial patient was well known. In 1880 Laveran, a French army surgeon in Algiers, began to study the blood of such patients microscopically and soon was able to demonstrate the parasite that caused the disease. His discoveries were not readily accepted, but other investigations soon confirmed his observations and the fact was gradually firmly established. Not until recently, however, did this distinguished physician receive a full recognition of his work. A few years ago he was awarded the Nobel prize for medicine, perhaps the highest honor that can be bestowed on any physician. It is interesting, too, to note in this connection that it was another French surgeon who in 1840 discovered that sulphate of quinine is a specific for malaria.

The next important step was made in 1885 by Golgi, an Italian, who studied the life-history of the parasite in the blood and distinguished the three forms which cause the three most familiar kinds of malarial fevers, the tertian, the quartan and the remittent types. From this time on this parasite has been studied by physicians of many nationalities and the whole course of its life-history worked out. In order that we may understand how it was that mosquitoes were determined to be the means of disseminating this parasite we will discuss first its life-history in the human blood.

The parasites that cause the malarial fevers are Sporozoans and belong to the genus Plasmodium. Other names such as Hamoeba and Laverania have been used for them, but the term Plasmodium is the one now most commonly employed. The three most common species are vivax, malari?and falciparum, causing respectively the tertian, quartan and remittent fevers.

LIFE-HISTORY OF PARASITE

The life-history of all of these is very similar, the principal difference being in the length of time it takes them to sporulate. Let us begin with the parasite after it has been introduced into the blood and trace its development there. At first it is slender and rod-like in shape. It has some power of movement in the blood-plasm. Very soon it attacks one of the red blood-corpuscles and gradually pierces its way through the wall and into the corpuscle substance (Fig. 99); here it becomes more amoeboid and continues to move about, feeding all the time on the corpuscle substance, gradually destroying the whole cell. As the parasite feeds and grows there is deposited within its body a blackish or brownish pigment known as melanin.

During the time that the parasite is feeding and growing it is also giving off waste products, as all living forms do in the process of metabolism, but as the parasite is completely inclosed in the corpuscle wall these waste products cannot escape until the wall bursts open. After about forty hours if the parasite is vivax or about sixty-five hours if it is malari?it becomes immobile, the nucleus divides again and again and the protoplasm collects around these nuclei, forming a number of small cells or spores, as they are called. In about forty-eight or seventy-two hours, depending on whether the parasite is vivax or malari?the wall of the corpuscle bursts and all these spores with the black pigment and the waste products that have been stored away within the cell are liberated into the blood-plasm.

These spores are round or somewhat amoeboid and are carried in the blood for a short time. Very soon, however, each one attacks a new red corpuscle and the process of feeding, growth and spore-formation continues, taking exactly the same time for development as in the first generation, so every forty-eight hours in the case of the vivax, and every seventy-two hours in the case of the malari?a new lot of these spores and the accompanying waste products are thrown out into the blood. Thus in a very short time many generations of this parasite occur and thousands or hundreds of thousands of the red-blood corpuscles are destroyed, leaving the patient weak and anemic. It will be seen, too, that the recurrence of the chills and fevers is simultaneous with the escaping of the parasites from the blood-corpuscles, together with the waste products of their metabolism.

These waste products are poisonous, and it is believed that this great amount of poison poured into the blood at one time causes the regular recurring crisis. Zoologists well know that this process of asexual reproduction, i. e., reproduction without any conjugation of two different cells, cannot go on indefinitely, and those who were studying the life-cycle of these parasites were at a loss to know where the sexual stage took place. In the meantime studies of other parasites more or less closely related to Plasmodium showed that the sexual stage occurred outside the vertebrate host. The remarkable work of Dr. Smith on the life-history of the germ that causes the Texas fever of cattle had a strong influence in directing the search for this other stage of the malarial parasite. Another thing that indicated that this sexual generation must take place outside the body of the vertebrate host was the fact that the investigators found that the parasites in certain of the cells did not sporulate as did the others. When these individuals were drawn from the circulation and placed on a slide for study it was found that they would swell up and free themselves from the inclosing corpuscle and some of them would emit long filaments which would dart away among the corpuscles.

Many men have worked on this problem, but perhaps the most credit for its solution will always be given to Sir Patrick Manson, the foremost authority on tropical diseases, and to Ronald Ross, a surgeon in the English army. There is no more interesting and inspiring reading than that which deals with the development of the hypothesis by Manson and the persistent faith of Ross in the correctness of this theory, and his continuous indefatigable labors in trying to demonstrate it. It was an important piece of scientific work, and shows what a man can do even when the obstacles seem insurmountable.

THE PARASITE IN THE MOSQUITO

Briefly stated again, the problem was this: We have here a parasite in the blood which behaves as do many other forms of life. Some of these parasites do not go on with their development until they are removed from the circulation. Now, how are they thus removed from the circulation under normal conditions? This must first be solved before the still greater and more important problem of how the parasite gets from one human host to another can be taken up. In studying this over Manson reasoned that certain suctorial insects were the agencies through which blood was most commonly removed from the circulation and he ventured the guess that this change in the parasite that may be seen taking place on the slide under the microscope, normally takes place in the stomach of some insect that sucks man's blood. Ross was greatly impressed with the theory and began his long and apparently hopeless task of finding these parasites in the stomach of some insect. When we remember that they are so minute that they can only be seen by the use of the highest power of the microscope we can realize something of the magnitude of the task. Ross, who was at that time stationed in India, selected the mosquito as the most likely of the insects to be the host that he was looking for. For over two and one-half years he worked with entirely negative results, for after examining thoroughly many thousands of mosquitoes he found no trace of the parasite.

Practically all his work was done on the most common mosquito of the region, a species of Culex. But one day a friend sent him a different mosquito, one with spotted wings, and in examining it he was interested to note certain oval or round nodules on the outer walls of the stomach. On closer examinations he found that each of these nodules contained a few granules of the coal-black melanin of malarial fever. Further studies and experiments showed that these particular cells could always be found in the walls of the stomach of this particular species of mosquito a few days after it had bitten a malarial patient. This epoch-making discovery was made in 1898. Ross was detailed by the English government to devote his whole time to the further solution of the problem, and after two years more of careful experimentation and study was able to give a complete life-history of this parasite. His experiments have been repeated many times, and the conclusions he arrived at are as undeniable as any of the known facts of science.

The whole life-history as we now know it can be summed up as follows: The parasites develop within the circulation but certain of them seem to wander about and do not go on with their development there. When these particular parasites are taken into the stomach of most mosquitoes they are digested with the rest of the blood. But when they are taken into the stomach of a mosquito belonging to the genus Anopheles or other closely related genera they are not digested but go on with their development, conjugation and fertilization taking place, resulting in a more elongated form which makes its way through the walls of the stomach on the outside of which are formed the little nodules discovered by Ross on his mosquitoes. Within these nodules further division and development takes place until finally the nodule is burst open and many thousand minute rod-like organisms, sporozoites, are turned loose into the body-cavity of the mosquito. Owing to some unknown cause these little organisms are gathered together in the large vacuolated cells of the salivary glands of the mosquito, and when the mosquito bites a man or any other animal they pour down through the ducts with the secretion and are thus again introduced in the circulation.

The nodules or cysts on the walls of the stomach of the mosquito may contain as many as ten thousand sporozoites, and as many as five hundred cysts may occur on a single stomach.

It takes ten, twelve or more days from the time the parasites are taken into the stomach of the mosquito before they can go through their transformations and reach the salivary gland, the time depending on the temperature. So it is ten or twelve days or sometimes as much as eighteen or twenty days from the time an Anopheles bites a malarial patient before it is dangerous or can spread the disease. On the other hand, the sporozoites may lie in the salivary gland alive and virulent for several weeks. It does not give up all the parasites at one time, so that three or four or more people may be affected by a single mosquito.

It is well known that two parasites may often be seen in the same corpuscle. This is often simply a case of multiple infection, but Dr. Craig has very recently shown that under certain conditions two individuals may enter the same corpuscle and conjugate and the resulting individual will be resistant to quinine and may remain latent in the spleen or bone marrow for a long time. Under favorable conditions it may again begin the process of multiplication and the patient will suffer a relapse.

SUMMARY

Now let us sum up some of the reasons why we believe that the malaria fever can be transmitted only through the agency of mosquitoes. First, we know the life-history of the parasite, it has been studied in both of its hosts. Attempts have been made to rear it in other hosts but without avail, and we know from the general relations of the parasite that it must have this sexual as well as the asexual generations. Second, in some regions which would seem to be malarial, that is, where the miasmatic mists arise, no malaria occurs. Why? Usually it can be definitely shown that no Anopheles occur there. Other mosquitoes may be there in abundance, but if no Anopheles, there is no malaria. In certain regions this is well demonstrated. The west coast of Africa is one of the worst pest-holes of malaria and Anopheles. The east coast has no malaria and no Anopheles. In many islands the same condition exists. On the other hand, the Fiji Islands have Anopheles but no malaria. No malaria has ever been introduced there to infect the mosquitoes. In the same way Stegomyia occurs in some of the South Sea islands and yet there is no yellow fever there.

EXPERIMENTS

We may review, too, a few of the classic experiments that have served to show that malaria can be contracted in no other way than through the bite of the mosquito.

For many years Grassi, an Italian, devoted almost his whole time to the study of malaria. In 1900 he received permission from the government to experiment on the employees of a piece of railroad that was being built through a malarial region. This was divided for the purpose of the experiment into three sections, a protected zone in the middle and an unprotected zone at each end.

Those working in the protected zone had their houses completely screened and no one was allowed out of doors after sunset except they were protected with veils and gloves. Early in the season they were all given doses of quinine to prevent auto-infection. In the unprotected zone no screens were used and every one was allowed to go without special protection. The result for the summer was that there were no new cases of fever in the protected zone. In the unprotected zones practically all had the fever as usual.

In the same year two English physicians, Sambon and Low, went to Italy where they built a cabin in one of the marshes noted as being a malaria pest-hole. The house was thoroughly screened so that no mosquitoes could enter, but the windows were always open so as to admit the air freely day and night. Here they lived for three months, out of doors as much as they pleased during the day but inside where they were protected from the mosquitoes at night. No quinine was used and no fever developed, although all about them other people were having the fever as usual.

Another English physician who had not been in malarial regions allowed himself to be bitten by infected mosquitoes sent from a malarial locality. In due time he developed the fever. Many other experiments made in various places might be cited. The results have all been practically the same. To-day the soldiers of many civilized nations are required to protect themselves from mosquitoes because it has been found that it pays. Disease has always been a worse terror than bullets in any war, and we are fast learning that the great loss from diseases heretofore considered unavoidable may be very largely eliminated by proper sanitary arrangements and protection from noxious insects.

CHAPTER VIII

MOSQUITOES AND YELLOW FEVER

Yellow fever is a disease, principally of seaport towns, from which the United States has suffered more than any other country. It is endemic only in tropical regions but is often carried to subtropical, sometimes even to temperate zones where, if the proper mosquitoes exist, it may rage until frost.

Vera Cruz, Havana, Rio de Janeiro, and the west coast of Africa were long regarded as permanent endemic foci, the disease appearing there in epidemic form from time to time, often spreading to other ports in more or less close communication with such places. In the United States the Gulf states have been the greatest sufferers from the disease, although it has spread as far as Baltimore, Philadelphia and Washington, where at rare intervals it was most serious, abating its ravages only when frost came.

The last severe outbreak occurred in New Orleans in 1905 when eight thousand cases and nine hundred deaths occurred. At that time there was waged one of the most remarkable warfares against death in its most terrifying form that the world has ever known. And, thanks to the achievements of science, particularly to the investigations of three men, one of whom gave his life to the cause, the fight was successful and this dreadful outbreak was checked just at the time when according to all precedent it should have been at its height.

This result which at other times and under other conditions would have been considered miraculous was achieved not by the usual custom of isolation, quarantine, etc., but by a direct, we may almost say hand to hand, conflict with mosquitoes: the mosquitoes belonging to a particular genus and species, Stegomyia calopus (fasciata).

Before taking up a discussion of this achievement in New Orleans let us consider first the work of the men that made such results possible.

For many years the cause and methods of dissemination of this disease had been a puzzle to physicians and scientists. Very early it was believed that it might be transmitted through the air, and the fact that infection usually occurred in the vicinity of the water and in the tropics or in midsummer led to the belief that the disease was due to fermentation. This theory received strong support in the fact that serious outbreaks of the fever often followed the coming into port of vessels from the tropics with the water in their holds in an offensive condition. When it was discovered that bacteria were the cause of fermentation and also of many diseases this theory was considered abundantly proven. From time to time, announcements have been made that the particular species of bacteria that causes the disease has been isolated, but there has always been something lacking in the final proof.

Yellow fever has always been regarded as a very highly contagious as well as infectious disease, and the utmost precaution has been taken to isolate the patients when possible and in recent years strict quarantines have been established against infected localities and no person or commerce or even the mails were allowed to come from such places without thorough fumigations. But all these things proved unsatisfactory. The disease could not ordinarily be checked by simply isolating the patients. Many people became sick without ever having been near a yellow fever patient, while others worked in direct daily contact with the disease and did not suffer from it. Those who had once had it and recovered became practically immune, rarely suffering from a second attack. Negroes may suffer from the disease, but are usually regarded as practically immune.

It was early observed, too, that the danger zone might be quite well defined and that outside this zone one would be safe. More than a century ago the British troops and other inhabitants of Jamaica found that by retreating to the mountains during the warm weather the non-immunes could escape the fever. It was also observed that those who slept on the first floor were more apt to take the disease than those on the second floor.

THE YELLOW FEVER COMMISSION

In 1900, during the American occupation of Cuba, yellow fever became very prevalent there. A board of medical officers was ordered to meet in Havana for the purpose of studying the disease under the favorable opportunities thus afforded. This board, which came to be known as the Yellow Fever Commission, was composed of Drs. Walter Reed, James Carroll, Jessie W. Lazear and Aristides Agramonte of the United States Army. Agramonte was a Cuban and an immune, the others were non-immunes. Dr. Manson in his lectures on Tropical Medicines says of them:

"I cannot pass on, however, to what I have to say in connection with this work without a word of admiration for the insight, the energy, the skill, the courage, and withal the modesty and simplicity of the leader of that remarkable band of workers. If any man deserved a monument to his memory, it was Reed. If any band of men deserve recognition at the hands of their countrymen, it is Reed's colleagues."

Their first work was to determine whether any of the germs that had been claimed to be the cause of yellow fever were really responsible for the disease. Bacillus icteroides that for some time and by some investigators had been named as the offender was particularly investigated, but was proved to be a secondary invader only.

Dr. Charles Finlay of Havana had been claiming for some years that the yellow fever was transmitted by means of the mosquito and possibly by other insects also. He even claimed to have proved this theory experimentally. We know now, however, that there must have been errors in his experiments and that his patients became infected from sources other than those he was dealing with.

The Yellow Fever Commission decided to put this theory to the test and secured a number of volunteers for the experiments. The first thing was to let an infected mosquito bite some non-immune person. How this was done and the results, may be told in Dr. Carroll's own words.

EXPERIMENTS

"Two separate lines of work now presented: one, the study of the bacterial flora of the intestine and anaerobic cultures from the blood and various organs; the other, the theory of the transmission of the disease by the mosquito, which had been advanced by Dr. Carlos Finlay in 1881. After due consideration it was decided to investigate the latter first. Then arose the question of the tremendous responsibility involved in the use of human beings for experimental purposes. It was concluded that the results themselves, if positive, would be sufficient justification of the undertaking. It was suggested that we subject ourselves to the same risk and this suggestion was accepted by Dr. Reed and Dr. Lazear. It became necessary for Dr. Reed to return to the United States and the work was begun by Dr. Lazear, who applied infected mosquitoes to a number of persons, himself included, without result. On the afternoon of July 27, 1900, I submitted myself to the bite of an infected mosquito applied by Dr. Lazear. The insect had been reared and hatched in the laboratory, had been caused to feed upon four cases of yellow fever, two of them severe, and two mild. The first patient, a severe case, was bitten twelve days before; the second, third and fourth patients had been bitten six, four and two days previously, and were in character mild, severe and mild respectively. In writing to Dr. Reed that night of the incident, I remarked jokingly that if there was anything in the mosquito theory, I should have a good dose. And so it happened. After having slight premonitory symptoms for two days, I was taken sick on August 31, and on September 1, I was carried to the yellow fever camp. My life was in the balance for three days, and my chart shows that on the fifth, sixth and seventh days my urine contained eighth-tenths and nine-tenths of moist albumin. On the day I was taken sick, August 31, 1900, Dr. Lazear applied the same mosquito, with three others, to another individual who suffered a comparatively mild attack and was well before I had left my bed. It so happened that I was the first person in whom the mosquito was proved to convey the disease.

"On the eighteenth of September, five days after I was permitted to leave my bed, Dr. Lazear was stricken, and died in convulsions just one week later, after several days of delirium with black vomit. Such is yellow fever.

"He was bitten by a stray mosquito while applying the other insects to a patient in one of the city hospitals. He did not recognize it as a Stegomyia, and thought it was a Culex. It was permitted to take its fill and he attached no importance to the bite until after he was taken sick, when he related the incident to me. I shall never forget the expression of alarm in his eyes when I last saw him alive in the third or fourth day of his illness. The spasmodic contractions of his diaphragm indicated that black vomit was impending, and he fully appreciated their significance. The dreaded vomit soon appeared. I was too weak to see him again in that condition, and there was nothing that I could do to help him.

"Dr. Lazear left a wife and two young children, one of whom he had never seen."

These experiments and many others like them conducted on soldiers and Spanish immigrants proved that this particular mosquito would transmit the disease under certain conditions.

1. The mosquito must bite the patient during the first three days of the fever; after that a yellow fever patient cannot infect a mosquito.

2. A period of twelve days must elapse before the mosquito is able to infect

another person. After that she may infect anyone she may bite; that is, the germs remain virulent during the rest of the mosquito's life. The French Yellow Fever Commission working in Rio de Janeiro claim that the first generation of offspring from such an infected mosquito is capable of causing the disease after they are fourteen days in the adult condition.

The next step was to ascertain whether the disease could be contracted in any other way than by the bites of infected mosquitoes. A camp named Camp Lazear was established and the following tests made: A mosquito-proof building of one room was completely divided by a wire screen from floor to ceiling. In one room fifteen mosquitoes that had previously bitten yellow fever patients and had undergone the proper period of incubation were liberated. In this room a non-immune exposed himself so that he was bitten by several of the insects. A little later the same day and again the next day the mosquitoes were allowed to feed on him for a few minutes. Five days later, the usual incubation period, he developed yellow fever.

At the same time that he entered the building two other non-immunes entered the other compartment where they slept for eighteen nights separated from the mosquitoes by the wire screen. They showed no signs of taking the fever.

In another mosquito-proof house two soldiers and a surgeon, all non-immunes, lived for twenty-one days. From time to time they were supplied with soiled articles of bedding, clothing, etc., direct from the yellow fever hospital in the city. These articles had been soiled by the urine, fecal matter and black vomit obtained from fatal and other cases of yellow fever. These articles were handled and shaken daily, but no disease developed among the men and at the end of the twenty-one days, two other non-immunes relieved them and handled a new supply of clothing in the same way, sleeping between the same sheets that had been used by a patient dying of yellow fever and exposing themselves in every possible way to the soiled clothing. But no disease developed. That these men were susceptible was shown later by inoculating some of them, when they developed the disease.

In another experiment certain men in a camp allowed themselves to be bitten by mosquitoes that had passed through the proper period of incubation and every one of them and no others contracted the disease. It was also shown that a mosquito was capable of communicating the disease as long as fifty-seven days after it had bitten a yellow fever patient. Another set of experiments showed that a subcutaneous injection into a non-immune of a very small quantity of blood from the veins of a yellow fever patient in the first two or three days of the disease would produce the fever.

SUMMARY OF RESULTS

Since that time much other work has been done by independent workers as well as by French and English Commissions both working at Rio de Janeiro. The results of their investigation are practically the same and may be summed up as follows:

1. The virus of the yellow fever is in the blood-plasma, not in the corpuscles, for these may be removed and the plasma still be infective.

2. The virus is conveyed from one patient to another by the yellow fever mosquito, Stegomyia calopus, and in no other way except by experimental injections.

3. The patient is a source of infection only during the first three or four days of the disease (this after the three to six days of incubation).

4. The virus must undergo an incubation period of twelve to fourteen days in the mosquito before she is capable of transmitting the disease.

5. The parasite, whatever it is, has never been seen. It is probably too small to be seen by any of our present microscopes, even the recently invented ultramicroscope. It is probably not a bacterial parasite but very likely a Protozoan, and certain specialists have even shown by the study of all the available data that it almost certainly belongs to the Sporozoan genus Spirocheta.

Now what does all this mean? It means the saving of hundreds of human lives annually. It means the banishing from many localities and possibly very soon from the face of the earth of a disease that since the earliest settlements on this continent has been a source of terror. It means the making habitable of certain places which heretofore a white man has entered only at the risk of his life. It means that quarantines need no longer be established when yellow fever breaks out in a district; quarantines which have inevitably caused the loss of millions of dollars to the world of commerce.

RESULTS IN HAVANA

The first practical work based on these findings was done in Havana. The Yellow Fever Commission made their recommendations in 1900. In 1901 and 1902 they were put into effect. The following table of the death rate there during a period of ten years shows graphically the results:

DEATHS IN HAVANA FROM YELLOW FEVER

	1893	1894	1895	1896	1897	1898	1899	1900	1901	1902
Jan.	15	7	15	10	69	7	1	8	7	0
Feb.	6	4	4	7	24	1	0	9	5	0
Mar.	4	2	2	3	30	2	1	4	1	0
Apr.	8	4	6	14	71	1	2	0	0	0
May	23	16	10	27	88	4	0	2	0	0
June	69	31	16	46	174	3	1	8	0	0
July	118	77	88	116	168	16	2	30	1	0
Aug.	100	73	120	262	102	16	13	49	2	0
Sep.	68	76	135	166	56	34	18	52	2	0
Oct.	46	40	102	240	42	26	25	74	0	0
Nov.	28	23	35	244	26	13	18	54	0	0
Dec.	11	29	20	147	8	13	22	20	0	0

As long as the United States held control at Havana the yellow fever was kept in check by fighting the mosquitoes, when this vigilance was relaxed the fever began to appear again and the Cubans found that it was necessary to keep up the fight against the mosquitoes if the island was to be kept free from the disease.

THE FIGHT IN NEW ORLEANS

In the summer of 1905 came another opportunity to put the knowledge gained during these experiments to a practical test. Samuel Hopkins Adams in his article in McClure's Magazine, June, 1906, says of the beginning of this fight:

"Eight years before, the mosquito-plague had infected the great, busy, joyous metropolis of the south. Ignorant of the real processes of the infection, New Orleans had fought it blindly, frantically, in an agony of panic, and when at last the frost put an end to the helpless city's plight, she lay spent and prostrate. The yellow fever of 1905 came with a more formidable and unexpected suddenness than that of 1897. It sprang into life like a secret and armed uprising in the midst of the city, full-fledged and terrible. But there arose against it the trained fighting line of scientific knowledge. Accepting, with a fine courage of faith that most important preventive discovery since vaccination, the mosquito dogma, the Crescent City marshaled her defenses. This time there was no panic, no mob-rule of terrified thousands, no mad rushing from stunned inertia to wildly impractical action; but instead the enlistment of the whole city in an army of sanitation. Every citizen became a soldier of the public health. And when, long before the plague-killing frost came, the battle was over, New Orleans had triumphed not only in the most brilliant hygienic victory ever achieved in America, but in a principle for which the whole nation owes her a debt of gratitude."

For some time the authorities had been trying to keep secret the fact that the disease was prevalent, but the rapidity with which it spread made them realize that only united action on the part of all the community would be of any avail. The Citizens Volunteer Ward Organizations were organized for the purpose of fighting the mosquitoes which were everywhere. To many the fight looked hopeless. The miles of open gutters, the thousands of cisterns and little pools of standing water everywhere furnished abundant breeding-places for the mosquitoes. The ditches and ponds were drained or salted, the cisterns were screened, infected houses were fumigated, yet the fever continued to spread. Rains refilled the ditches, winds tore the screens from the cisterns, the ignorant people of the French quarter refused to cooperate. At last the city in desperation appealed to the President for aid. Surgeon J.H. White and a number of officers and men of the United States Public Health and Marine Hospital Service soon took charge of the work. This was continued along the same lines as before with the same object in view. But with the coming of the regulars the work was more systematically and thoroughly done. Every case of fever was treated as though it was yellow fever and every precaution taken to prevent mosquitoes from biting such a patient. The houses in which the fever occurred were thoroughly fumigated to kill any mosquitoes that might be there, and the neighborhood was thoroughly searched to find any places where the mosquitoes might be breeding. So confident were the authorities that the mosquito was the sole cause of the disease spreading, that besides fighting it no other work was undertaken save to make the sick as comfortable as possible.

Finally the results began to be apparent. The number of cases gradually diminished, until long before frost came the city was free from the great pest. Yellow fever will doubtless appear from time to time in New Orleans and other cities, but there is, at least there should be, small danger of another great epidemic, for the people now know how the disease is caused and the remedy.

Not long since I had occasion to write to a prominent entomologist in Louisiana for some specimens of the yellow fever mosquito for laboratory work. The following extract from his reply will show something of the work that is still being done there.

"I am afraid we cannot furnish specimens of Stegomyia, in spite of the fact that Louisiana is supposed to be the most favorable home of this species in the South. Since the light occurrence of yellow fever in this State in 1905, a very vigorous war has been kept up against Stegomyia, and the ordinances of all Louisiana cities and principal towns require the draining of all breeding places of this mosquito and the constant oiling or screening of all cisterns or other water containers. The result is this species is very rare. Here in Baton Rouge I only see one once in a great while, and it would require perhaps a good many days' work at the present season to get as good specimens and as many of them as you require."

IN THE PANAMA CANAL ZONE

Yellow fever was one of the worst obstacles that confronted the French when they were attempting to build the Panama Canal. The story of the suffering and death from this dread disease there is most pathetic. Ship-load after ship-load of laborers were sent over, as those who had gone earlier succumbed to the fever. The contractors were responsible for their men while they were sick and in order to avoid having to pay hospital expenses the men were often discharged as soon as they showed signs of sickness.

Many of them died along the roadside while endeavoring to reach some place where they could obtain aid. The hospitals were usually filled with yellow fever patients, a very large percentage of whom died.

Not only the day laborers suffered but many of the engineers, doctors, nurses and others sickened and died of the disease. It is reported that eighteen young French engineers came over on one vessel and in a month after their arrival all but one had died of the yellow fever. Out of thirty-six nurses brought over at one time, twenty-four died of the fever, and during one month nine members of the medical staff of one of the hospitals succumbed.

One of the first things that the United States Government did in beginning work in the canal zone was to take up the fight against the yellow fever mosquito. In Panama where the water for domestic purposes was kept in cisterns and water-barrels, inspectors were appointed to see that all such receptacles and other possible breeding-places for mosquitoes were kept covered. After the first inspection, 4,000 breeding-places were reported. About six months later there were less than 400. Similar work was done in all the towns and settlements along the route of the canal. In addition to this fight against the yellow fever mosquito considerable attention was paid to the breeding-places of the malarial mosquito. The results have been remarkable. Cases of yellow fever are now rare throughout this zone, and there has been a very great reduction in the extent of the malarial districts. The last case of yellow fever occurred in May, 1906. Before this work was done a man took his life in his hands when he went into this region. Now it is regarded as a perfectly safe place to live. Indeed it is a much safer place than many sections of our own country where proper sanitary measures have not been taken to protect the health of the community.

IN RIO DE JANEIRO

In Rio de Janeiro they have as yet been unable to get rid of the mosquitoes, although thousands of dollars are spent annually in fighting them. But the non-immunes there protect themselves by doing their business in Rio during the day and going back at night to Petropolis, twenty-five miles inland and twenty-five hundred feet higher, where they are safe, for no Stegomyia have ever been found there.

They claim there that the yellow fever mosquito does not bite during the daytime after she has laid her eggs, and that she will not lay her eggs until about three days after she has fed on blood, therefore a Stegomyia that bites during the day will not carry the yellow fever because she is too young. This seems to explain why the fever cannot be contracted by being bitten by a mosquito in the daytime. Certain other experiments, however, have given different results so that as far as we know it is not safe to be bitten at any time by such a mosquito in a region where the disease is endemic or where it is epidemic.

In the main the work of the French Yellow Fever Commission working in Rio de Janeiro has confirmed the findings of the American Commission. One interesting special thing that the French Commission seems to have established is that the female may transmit the infecting power to her offspring, so that it would be possible for a mosquito that had never bitten a yellow fever patient to be capable of infecting a non-immune person. While all this is very probable in the light of what we know of the disease and the way in which other diseases caused by similar organisms may be transmitted by the parent to the offspring, yet the most conservative investigators are waiting for further proof.

HABITS OF STEGOMYIA

The whole fight against yellow fever, then is directed, as we have seen, against the mosquito, Stegomyia calopus. The habits of this species are such as to make it easy in some respects to combat. It is seldom found far away from human habitation. The adults will not fly far. Once in a house they usually stay there except when they leave to deposit their eggs.

On the other hand, some of these same habits make it all the more dangerous. It will breed in almost any kind of water, no matter how filthy, and a very small amount will suffice. Thus any leaks from water-pipes or drains, cisterns, small cans of water or any such places may become dangerous breeding-places. If conditions are unfavorable there will often be developed small individuals which can easily make their way through ordinary mosquito-netting.

Dr. Manson has pointed out an interesting possible result of the crusade that is now being waged against the yellow fever mosquitoes. The immunity of the people native to the endemic regions is supposed to be due to their having had mild attacks of the fever during childhood, for the children in these regions are subject to certain fevers which are probably very mild forms of yellow fever.

Now if we kill practically all of the Stegomyia so that these children do not have this fever there will be developed, in due time, a population most of whom are non-immune.

This freedom from the disease for some time will allow us to grow careless in regard to fighting the mosquitoes. They will be allowed to increase and by some chance the yellow fever will again be introduced and there will then be very grave danger of most extensive and destructive epidemics.

DANGER OF THE DISEASE IN THE PACIFIC ISLANDS

I have already referred once or twice to the conditions in many of the Pacific tropical islands. In some of these various species of Stegomyia are abundant, and in some Stegomyia calopus is the most abundant and troublesome form. All the natives of these islands are non-immune because there has never been any yellow fever there. Unless extraordinary care is taken the disease will be introduced there sooner or later and the results are sure to be most appalling. The climatic and sanitary conditions and the habits of the people are ideal for the development and spread of the disease, and what I have seen of the conditions on some of these islands convinces me that it would be almost impossible to control the disease before it had a chance to kill a large percentage of the population.

With the opening of the Panama Canal these things become more possible. Heretofore, the shipping to these regions has not been from ports where yellow fever was endemic or even likely to be epidemic. But unless the yellow fever is kept out of the canal zone, the danger will be many fold what it is now.

The white man has already carried enough misery to these island peoples in the way of loathsome diseases, and it is to be hoped that this, another great curse, will not be carried to them with our civilization, the beneficial results of which have been so often very justly questioned.

What I have said in regard to these islands applies with equal force and in some instances with even greater force to parts of Asia, the Eastern Archipelago and other places.

CHAPTER IX

FLEAS AND PLAGUE

Plague has always been one of the most dreaded diseases, and when we read of its ravages in the old world and the utter helplessness of the people before it we do not wonder that the very word filled them with horror. One of the greatest scourges ever known began in Egypt about A.D. 542, and spread along the shores of the Mediterranean to Europe and Asia. It lasted for sixty years, appearing again and again in the same place and decimating whole communities.

Another great pandemic, beginning in 1364, spread over the whole of the then known world and appeared in its most virulent form. On account of diffuse subcutaneous hemorrhages it came to be known as the "black death" and of course spread terror in all the communities where it appeared. Whole villages and districts were depopulated. The death-rate was very high, one authority placing the total mortality at twenty-five million.

During this time new centers of infection were established, and since then it has been carried by the commerce of the nations to all parts of the world. It is not restricted, as many other epidemic diseases, to the tropics or semi-tropics, although as a matter of fact we find it is more prevalent in these regions on account of the sanitary conditions.

HOW PLAGUE WAS CONTROLLED IN SAN FRANCISCO

Attention is called to these things in order that we may compare past conditions with present. During the last few years San Francisco has been fighting an outbreak of plague that in other days would have been nothing less than a national calamity. But with modern methods of handling it, based on knowing what it is, what causes it and how it is spread, the authorities there have been able not only to hold the disease in check, but practically to stamp it out with the loss of comparatively few lives.

Dr. Blue of the Public Health and Marine Hospital Service and his co-workers directed their whole energy toward controlling the rats. A small army of men were employed, catching rats in every quarter of the city. Dr. Rucker reports that fully a million rats were slain in this campaign. Their breeding-places were destroyed by making cellars, woodsheds, warehouses, etc., rat-proof and removing all old rubbish. Garbage cans were installed in all parts of the city, as it was required that all garbage be stored where rats could not feed upon it, and altogether every effort was made to make it as uncomfortable as possible for the rats.

The marked success attending this work abundantly confirms the soundness of the theory upon which it was based, and serves as another example of the way in which science is teaching us how to prevent or control many of our most serious diseases.

THE INDIAN PLAGUE COMMISSION

In 1896, what proved to be a very serious outbreak of plague, occurred in Bombay and spread to other parts of India. In 1898, a commission was appointed to inquire into the origin of the different outbreaks, the manner in which the disease is communicated, etc. This was known as the Indian Plague Commission, and its exhaustive report, together with the minutes of the evidence presented to the committee, represents a stupendous amount of work on this subject and is the basis for much of the later investigation that has been undertaken.

After the consideration of the evidence from various sources the commission decided that the principal mode of infection both for man and rats was through some sort of an abrasion in the skin, although it recognized also the possibility of infection through the nose and throat, and possibly, very rarely, through the intestinal tract or other places.

Considerable time was spent in considering Dr. Simond's claim, made in 1898, that fleas which have been parasitic on plague-infected rats migrate on the death of their hosts and convey the infection to healthy men and rats. Dr. Simond sought to establish the following:

"Firstly, that plague rats are eminently infective when infected with fleas and that they cease to be infective when they have been deserted by their parasites: Secondly, that living plague bacilli are found in association with fleas which are taken from plague-infected rats: Thirdly, that plague can pass from infected rats to other animals which have not come directly in contact with them or with their infected excretions: Fourthly, that fleas which infest rats will transfer themselves as parasites to men."

After reviewing the experiments which had been made to establish these claims the commission believed that sufficient precaution had not been taken to prevent infection from other sources and that not enough definite evidence was produced. Against this claim much negative evidence was considered and the final conclusion was "that suctorial insects do not come under consideration in connection with the spread of plague."

In 1905 another body of men known as the Advisory Committee was appointed to arrange for further studies in India and other places, particularly in relation to the mode of dissemination of the disease. They at once appointed a new working commission who immediately began their studies and experiments. The preliminary reports of their work, which are still known as the Reports of the Indian Plague Commission, as well as the reports of contributing investigations that are being made from time to time, have served to establish entirely Dr. Simond's claims and have completely revolutionized the methods of fighting plague.

There are several different types of plague, seeming to depend largely on the manner of infection. The most common type is that known as the bubonic plague which is characterized by buboes or swellings in various parts of the body. This form of infection is usually received through the skin in some manner or other. Only rarely does direct man-to-man infection occur though there is always the possibility of it. The investigations have shown that the flea is the most common agent in transferring the disease from rat to rat or from rat to man. This may be accomplished by the flea transferring the bacilli directly from one host to another on its proboscis, or they may be carried in the alimentary canal of the flea and gain an entrance into the skin through an abrasion of some kind when the flea is crushed as it is biting, or when some of the bacilli are left on the skin in the excreta of the insect.

RESULTS OF VERJBITSKI'S EXPERIMENTS

A very important series of experiments bearing directly on this subject was made in 1902 and 1903 by Dr. D.T. Verjbitski. The paper giving the results of this work was not published in any scientific journal until 1908 when the Advisory Committee published it in one of their reports. The experiments were so well planned and executed and the results so definite that I think it is worth while to give in full his summary of results. The bugs referred to are bedbugs.

"(1) All fleas and bugs which have sucked the blood of animals dying from plague contain plague microbes.

"(2) Fleas and bugs which have sucked the blood of animals which are suffering from plague only contain plague microbes when the bites have been inflicted from 12 to 26 hours before the death of the animals, that is, during that period of their illness when their blood contains plague bacilli.

"(3) The vitality and virulence of the plague microbes are preserved in these insects.

"(4) Plague bacilli may be found in fleas from four to six days after they have sucked the blood of an animal dying with plague. In bugs, not previously starved or starved only for a short time (one to seven days), the plague microbes disappear on the third day; in those that have been starved for four to four and one-half months, after eight or nine days.

"(5) The numbers of plague microbes in the infected fleas and bugs increase during the first few days.

"(6) The infected fleas and bugs contain virulent plague microbes as long as they persist in the alimentary canal of these insects.

"(7) Animals could not be infected by the bites of fleas and bugs which had been infected by animals whose own infection had been occasioned by a culture of small virulence, notwithstanding the fact that the insects may be found to contain abundant plague microbes.

"(8) Fleas and bugs that have fed upon animals which have been infected by cultures of high virulence convey infection by means of bites, and the more certainly so the more virulent the culture with which the first animal was inoculated.

"(9) The local inflammatory reaction in animals which have died from plague occasioned by the bites of infected insects is either very slight or absent. In the latter case it is only by the situation of the primary bubo that one can approximately identify the area through which the plague infection entered the organism.

"(10) Infected fleas communicate the disease to healthy animals for three days after infection. Infected bugs have the power of doing so for five days.

"(11) It was not found possible for more than two animals to be infected by the bites of the same bugs.

"(12) The crushing of infected bugs in situ during the process of biting, occasioned in the majority of cases the infection of the healthy animal with plague.

"(13) The injury to the skin occasioned by the bite of bugs or fleas offers a channel through which the plague microbes can easily enter the body and occasion death from plague.

"(14) Crushed infected bugs and fleas and their f鍢es, like other plague material, can infect through the small punctures of the skin caused by the bites of bugs and fleas, but only for a short time after the infliction of these bites.

"(15) In the case of linen and other fabrics soiled by crushing infected fleas and bugs on them, or by the f鍢es of these insects the plague microbes can under favorable conditions remain alive and virulent during more than five months.

"(16) Chemical disinfectants do not in the ordinary course of application kill plague microbes in infected fleas and bugs.

"(17) The rat flea Typhlopsylla musculi does not bite human beings.

"(18) Human fleas do bite rats.

"(19) Fleas found on dogs and cats bite both human beings and rats.

"(20) Human fleas and fleas found on cats and dogs can live on rats as casual parasites, and therefore can under certain conditions play a part in the transmission of plague from rats to human beings, and vice versa."

RESULTS OF VARIOUS INVESTIGATIONS

Various other plague commissions from other countries as well as many individuals have investigated the same subject, and the results all point conclusively to the fact that the rats and the fleas are at least the most important factors in the spread of the disease. The evidence from many sources and from many experiments may be briefly summed up as follows: The disease is caused by the presence in the system of minute bacteria, Bacillus pestis. It is probable that plague is primarily a disease of rats and only secondarily and accidentally, as it were, a disease of man.

Rats are subject to the plague and are often killed by it in great numbers. An outbreak of plague among men is often preceded by a very noticeable outbreak among rats.

Rats dying of the plague have their blood filled with the plague bacillus. Fleas or other suctorial insects feeding on such rats take myriads of these bacilli into their stomach and get many on their proboscis.

The fleas usually leave a rat as soon as it dies and of course seek some other source of food. When such infected fleas are permitted to bite other rats or guinea-pigs these animals often develop the disease. Several of the species of fleas that infest rats will bite man also, and in the cases of many plague patients it can be definitely shown that they had recently been bitten by fleas.

STRUCTURE AND HABITS OF FLEAS

A study of the structure and habits of fleas shows that in many respects they are particularly adapted for spreading such a disease as bubonic plague. The piercing proboscis consists of three long needle-like organs, the epipharynx and mandibles, and a lower lip or labium. The mandibles have the sides serrate like a two-edged saw. The labium is divided close to its base so that it really consists of two slender four-segmented organs which lie close together and form a groove in which the piercing organs lie. When the flea is feeding, the epipharynx and mandibles are thrust into the skin of the victim, the labium serving as a guide. As the sharp cutting organs are thrust deeper and deeper the labium doubles back like a bow and does not enter the skin. Saliva is then poured into the wound through minute grooves in the mandibles, and the blood is sucked up into the mouth by the sucking organ which lies in the head at the base of the mouth-parts. Just above this piercing proboscis is a pair of flat, obtuse, somewhat triangular pieces, the maxillary blades or maxill? When the proboscis is fully inserted into the skin the tips of these maxill?may also be embedded in the tissue and perhaps help to make the wound larger. Attached to these maxill?is a pair of rather stout, four-jointed appendages, the palpi. They probably act as feelers.

If the flea chances to be feeding on a plague-infected rat or person many of the plague bacilli will get on the mouth-parts and myriads of them are of course sucked up into the stomach with the blood. Those on the proboscis may be transferred directly to the next victim that it is thrust into, and those in the stomach may be carried for some time and finally liberated when the flea is feeding again or when it is crushed by the annoyed host. The latter is probably the most common method of infection, for the bacilli that are liberated when the flea is crushed may readily be rubbed into the wound made by the flea bite or into abrasions of the skin due to the scratching. Kill the flea, but don't "rub it in."

During the recent outbreak in San Francisco many thousand fleas that were infesting man, rats, mice, cats, and dogs, squirrels and other animals have been studied and it has been found that while each flea species has its particular host upon which it is principally found, few if any of them will hesitate to leave this host when it is dead and attack man or any other animal that may be convenient.

COMMON SPECIES OF FLEAS

Throughout India and in all the warm climates where plague frequently occurs the most common flea found on rats has come to be known as the plague flea (L 鎚 opsylla cheopus) (Figs. 105, 106), and is doubtless the principal species that is concerned in carrying the disease in those climates. It now occurs quite commonly on the rats in the San Francisco Bay region and is occasionally found there on man also. In the United States, Great Britain and other temperate regions another larger species, Ceratophyllus fasciatus is by far the most common flea found on rats, and is commonly known as the rat flea. It occurs on both the brown and the black rats Mus norvegicus and M. rattus, on the house mouse and frequently on man. It has also been taken in California on pocket gophers and on a skunk.

The common human flea (Pulex irritans) (Figs. 108, 109), is found in all parts of the inhabited world. Although we regard it primarily as a pest of human beings it often occurs very abundantly on cats, dogs, mice and rats as well as on some wild mammals such as badgers, foxes and others and has occasionally been found on birds.

Most entomologists regard the fleas commonly found on cats and dogs as belonging to one species Ctenocephalus canis. Others believe them to be distinct species and call the cat flea Ctenocephalus felis. So far as our personal comfort and safety is concerned it makes but little difference to us whether the flea that bites us is called canis or felis for they both look very much alike, and act alike and the bite of one hurts just as much as the bite of the other. Although cats and dogs are their normal hosts they are very often troublesome household pests, sometimes making a house almost uninhabitable. They are frequently found on rats, and therefore may carry the plague bacillus from rat to rat or from rat to man.

GROUND-SQUIRRELS AND PLAGUE

As early as 1903 Dr. Blue, in charge of the plague suppressive measures in San Francisco, became impressed with the possibility of the common California ground-squirrels (Otospermophilus beecheyi), acting as an agent in the transmission of plague. It was rumored at that time that some epidemic disease was killing the squirrels in some of the counties surrounding San Francisco Bay, notably in Contra Costa County. None of the squirrels were examined at that time, but since then many thousand have been carefully studied and it has been definitely shown that many of them are plague-infected. Just how the plague got started among them will probably never be really known. There is little doubt, however, but that it was transferred in some way from the rats to the squirrels. The trains and the bay and river steamers running out from San Francisco would afford abundant opportunity for the rats to go from the city to the warehouses all along the shore. Once there they would use the same runways as the squirrels about the warehouses and in the near-by fields. In harvest time the rats migrate to the fields and make constant use of the squirrel holes. The farmers in some sections report that they frequently catch more rats than squirrels in traps set in squirrel holes at that season of the year.

This close association of the rats and the squirrels affords a good opportunity for the fleas infesting them to pass from one host to the other.

So far only two species of fleas have been recorded from the ground-squirrels. One, Ceratophyllus acutus, is very common, sometimes literally swarming over the squirrels, particularly if a squirrel is sick or weak from any cause. The other species, Hoplopsyllus anomalus, is less abundant but still quite common. Both of these species infest rats also, so the chain of evidence is practically complete. We have only to assume that at sometime one or more of the plague-infected rats found their way into the region where the squirrels were, and the fleas passing from the rats to the squirrels would carry the plague with them.

The fact that the plague already has such a start among the squirrels opens a new and very serious phase of the problem of suppressing the disease. All who have hunted the ground-squirrels will testify to the readiness with which the fleas from them will bite those who are handling them. As it is the sick or weak squirrels that are most often taken there is always a chance that plague may be transferred from them to human beings. The records of the plague cases in California show at least three cases in which there seems to be very little doubt that the disease resulted from handling plague-infected squirrels.

A still more serious thing is the possibility of the disease remaining in a more or less virulent form among the squirrels for some time, possibly for years, and then breaking out again in some locality where the rats or men may become infected. As long as there is a trace of the disease among the squirrels there is always the chance of it spreading, so that new areas may become infested. Those in charge of the plague-suppressive measures are fully aware of these dangers and are making a careful study of the situation and will doubtless be able to cope with it successfully. It may be that the squirrels will have to be exterminated in the infected regions. This would be a long and difficult task, but the success attending the fight against the rats in a great city shows what can be done when the determination to do it is there.

REMEDIES FOR FLEAS

We have seen how a great city set to work to rid itself of the plague-sick rats. As a matter of fact it was not the rats that they were after primarily. If the rats had not harbored fleas the city would have been glad to let the disease take its course and destroy as many rats as possible. But it was found that the only way to get rid of the fleas that might possibly be infected with the plague was to kill their rat hosts.

General cleaning-up measures will of course very materially lessen the number of fleas about the private dwellings, but there often remains a number of fleas in the house that are a source of great annoyance even if the danger is eliminated.

Particularly is this apt to be so in places where cats or dogs are members of the household. These animals almost always harbor at least a few fleas, and where there are a few there is always a possibility, even a great probability, that there will be many more unless an effort is made to get rid of them.

In some sections of the country it is the cat and dog flea that is the most troublesome to man. The minute white eggs of the fleas are usually laid about the sleeping-places of these animals and the slender active larvaethat hatch from them feed upon any kind of organic matter that they can find in the dust or in the cracks and crevices. About eight or ten days after hatching the larvaespin delicate brownish cocoons in which they pass the pupal stage, issuing a few days later as the adult fleas.

It will at once appear, then, that it is important to provide the cats and dogs with sleeping-places that can be kept clean. If they have a mat or blanket to sleep on this can be taken up and shaken frequently and the dust swept up and burned. In this way many of the eggs or larvaemay be destroyed. Very often the dust under a carpet that has not been taken up and dusted for some time will be found to be harboring a multitude of fleas or their larvae In such cases a thorough cleaning of the carpet and the floors will bring relief. Houses that are unused for some time during the summer months are often found to be overrun with fleas in the fall, for the fleas have had an unmolested opportunity to breed and multiply. Such rooms of course require a thorough cleaning or it is sometimes possible to kill the fleas by a liberal use of pyrethrum powder or benzine or to fumigate. In this connection, Dr. Skinner's note in the Journal of Economic Entomology is worth repeating.

"In the latter part of last May (1908) I moved into a house that had not been previously occupied. No carpet was used and being summer only a few rugs were placed on the floors. A part of the household consisted of a collie dog and three Persian cats. Very soon the fleas appeared, the dog and cat flea, Ctenocephalus canis. I did not count them and I can't say whether they numbered a million or only a hundred thousand. On arising in the morning and stepping on the floor one would find from three to a dozen on the ankles. The usual remedies for fleas are either drastic or somewhat unsatisfactory. The drastic one is to send the animals to the institutions, where they are asphyxiated, or take the other advice, 'Don't keep animals.'

"I tried mopping the floors with rather a strong solution of creolin but it did little good. Previous experience with pyrethrum was not very satisfactory. Knowing the volatility of naphthalene in warm weather and the irritating character of its vapor led me to try it. I took one room at a time, scattered on the floor five pounds of flake naphthalene and closed it for twenty-four hours. On entering such a room the naphthalene vapor will instantly bring tears to the eyes and cause coughing and irritation of the air passages. I mention this to show how it acts on the fleas. It proved to be a perfect and effectual remedy and very inexpensive, as the naphthalene could be swept up and transferred to other rooms. So far as I am concerned the flea question is solved and if I have further trouble I know the remedy. I intend to keep the dog and the cats."

CHAPTER X

OTHER DISEASES, MOSTLY TROPICAL, KNOWN OR THOUGHT TO BE TRANSMITTED BY INSECTS

SLEEPING SICKNESS

One of the worst scourges of Africa and one that is to-day attracting world-wide attention is the disease known as trypanosomiasis, the terminal phase of which is sleeping sickness, one of the most ghastly diseases that we know.

Among the Protozoa referred to in one of the earlier chapters mention was made of certain trypanosomes which inhabit the blood of man and certain animals. Very little was known concerning these parasites previous to the beginning of the present century, but since that time several have been found to be of great economic importance. The group is being studied extensively and every day our knowledge of them is increasing so that we now know quite definitely the life-history of several.

Trypanosoma lewisi, a parasite of rats, is perhaps the best known as it is always common where-ever rats are found. Sometimes as many as 30% or 40% of the rats of certain districts are infected. It is thought that these are transmitted from rat to rat by the common rat-louse which serves as an intermediate host. Fleas may also act as disseminating agents.

A few other kinds cause serious disease of animals, but we are more interested just now in the particular one that is causing so much trouble in Africa. This parasite was discovered in 1902 and was named Trypanosoma gambiensi (Fig. 111). Since then it has been found to be widely distributed. Although the natives have doubtless long been subject to the disease caused by this parasite, the recent influx of whites to these regions and the consequent movements of the natives have caused a great spread of the disease so that whole regions are now made desolate, the inhabitants dying or fleeing to escape the uncanny death.

The disease may run its course in a few months or it may take years. The symptoms are various, but infection is usually soon followed by fevers, sometimes mild, sometimes severe, which recur at irregular intervals. Certain glands or other parts of the body may become swollen. More or less extensive skin eruptions occur on all parts of the body and the patient gradually becomes anemic and physically and intellectually feeble. The nervous system seems to be affected by the parasite, either directly or by the action of the toxins it produces. The patient becomes more debilitated and morose with an increasing tendency to sleep, hence the name sleeping sickness. As the stupor deepens the patient looses all desire or power of exertion and as little food is taken he rapidly wastes away and finally succumbs for after this final stage is reached there is no relief.

It is definitely known that a species of tsetse-fly, Glossina palpalis (Fig. 112), which somewhat resembles our stable-fly, is responsible for the dissemination of the disease, and some recent investigators have suggested that certain species of mosquitoes may also carry the parasite from one host to another. There still remains some doubt as to the exact manner in which the fly transmits the disease, but it seems altogether likely that it is an alternative host and does not serve as a simple mechanical carrier. In this respect it is like the mosquito which is one of the necessary hosts of the malaria parasites, and unlike the house-fly which carries the germs of various diseases in a purely mechanical way without serving as a definite necessary host for the parasite.

The tsetse-fly is found only in tropical Africa and is limited in its distribution there to certain very definite, narrow, brushy areas along the water's edge. If these places can be avoided there seems to be little danger. Those who are fighting the disease have found that if the brush in the vicinity of watering-places and ferry-landings is cleared away such places become comparatively safe. These flies do not lay eggs but produce full-grown larvaewhich soon pupate in the ground.

ELEPHANTIASIS

In many tropical regions human blood as well as that of other animals is the normal habitat of certain worm-like parasites (Nematodes). They are not entirely confined to the tropics but may extend far up into the subtropical regions. Five or six different species of these parasites are known, only one of which, however, has been shown to be of any pathological importance, as far as human beings are concerned.

[Illustration: FIG. 111--Trypanosoma gambiense; various forms from blood and cerebrospinal fluid. (After Manson.)]

[Illustration: FIG. 112--Tsetse-fly. (After Manson.)]

This species, Filaria bancrofti, is not only very widely distributed, but in regions such as some of the South Sea Islands a very large per cent of the natives have the filari?present in their blood. When these parasites are withdrawn from the circulation and placed on a slide for study they are seen to be minute transparent, colorless, snake-like organisms inclosed in a very delicate sack or sheath. They are but a little more than one-hundredth of an inch long and about as big around as a red blood-corpuscle. These are the larval forms of the parasite and have been called by Le Dantec the micro-filaria.

If blood of the patient drawn from the skin, is examined during the day few if any of these parasites are found, but if it is examined between five or six o'clock in the evening and eight or nine o'clock the next morning they may be found in numbers. During the daytime they have retired from the peripheral circulation to the larger arteries and to the lungs, where they may be found in great numbers.

This night-swarming to the peripheral circulation has been found to be a remarkable adaptation in the life-history of the parasite, for it has been demonstrated that in order to go on with its development these larval forms must be taken into the alimentary canal of the mosquito. Most of the mosquitoes in which the development takes place are night-feeders, so that the parasites are sucked up with the blood of the victim. Once inside the stomach they soon free themselves from the inclosing sheath and make their way through the walls of the stomach and enter the muscular tissue, particularly the thoracic muscles. Here they undergo a metamorphosis and increase enormously in size, some attaining one-sixteenth of an inch in length.

After sixteen to twenty days they leave these muscles and make their way to other parts of the body. A few may be found in different parts of the abdomen, but most of them make their way forward into the head of the mosquito and coil themselves up close to the base of the proboscis, finally finding their way down into the proboscis inside the labium. Here they lie until an opportunity offers for them to escape to the warm blood of a vertebrate. They probably pass through the thin membrane connecting the labella with the proboscis and there find their way into the wound made by the puncture when the insect bites. Whether these parasites can gain an entrance into the circulatory system in any other way is not known. It has been suggested that the mosquitoes dying and disintegrating on the surface of water may liberate the filari?which may later find their way into the system of the vertebrate host when the water is used for drinking, but most of the investigations made so far seem to indicate that they make their way directly from the proboscis into the new host.

Soon after entering the circulatory system of the human host the parasites make their way into the lymphatics where they attain sexual maturity, and in due time new generations of the larval filari?or microfilari?are poured into the lymph, and finally into the definite blood-vessels, ready to be sucked up by the next mosquito that feeds on the patient.

In most cases of infection the presence of these filari?in the blood seems to cause no inconvenience to the host. They are probably never injurious in the larval stage, that is, in the stage in which they are found in the peripheral circulation.

In many cases, however, the presence of the sexual forms in the lymphatics may cause serious complications. The most common of these is that hideous and loathsome disease known as elephantiasis in which certain parts of the patient becomes greatly swollen and distorted. An arm or a leg may become swollen to several times its natural size, or other parts of the body may be seriously affected.

In some of the South Sea Islands 30% to 40% of the natives are afflicted in this way, some only slightly others seriously. There is little or no pain, but in severe cases the distorted parts often render the patient entirely helpless.

The exact way in which the parasites cause such swelling is not very definitely known. Manson, who has done more work on these diseases than any one else, believes that the trouble arises from the clogging of the lymphatic glands or trunks, thus cutting them off from the general circulation, in which case the affected parts may become distorted. This clogging of the passages is believed to be due to the presence of great numbers of immature eggs which have been liberated by parasites injured in some way before their eggs were entirely developed.

This interference with the lymphatic circulation brings about the anomalous condition of a patient with a serious filarial disease with fewer of the filarial parasites in his blood than one who is not so seriously affected. This is supposed to be due to the fact that the disease-producing parasites have died and that the lymphatics have become so obstructed that any microfilari?they may contain cannot make their way into the general circulation. Such a patient then would not be as likely to infect a mosquito as would one less seriously affected.

It has always been thought that little or nothing could be done in the way of successfully treating this disease, but quite recently a French physician, who has been conducting a long series of experiments in the Society Islands, announced that he is able to cure many cases by certain surgical operations on the affected parts.

DENGUE OR "BREAKBONE FEVER"

This is another disease of the tropics often occurring in widespread epidemics. It is probably most frequently met with in the West Indies, but may occur in any of the tropical countries or islands. Occasionally it spreads into subtropical or even temperate regions. Several extensive epidemics have occurred in the United States. Once introduced into a community it spreads very rapidly and nothing seems to confer immunity.

The various names by which it has been called well describe its effect on the patient; breakbone fever, dandy-fever, stiff-necked or giraffe-fever, boquet (or "bucket") fever, scarlatina rheumatica, polka-fever, etc. While the suffering is intense as long as the disease lasts it seldom terminates fatally.

It has always been classed as a very contagious disease and it has not yet been definitely shown that it is not. Recent observations, however, have shown that it is probably caused by a certain Protozoan parasite that is found in the blood of dengue patients and several experiments have been conducted by Dr. Graham which seem to indicate that it is transmitted by mosquitoes. In these experiments, Culex fatigans, a common tropical or subtropical mosquito, was used. The same parasite that is found in the human blood may be found in the stomach and blood of the mosquitoes up to the fifth day after it has fed on a dengue patient.

Sick and healthy individuals were allowed to remain in close contact in a room from which the mosquitoes had been excluded, and the disease was not spread. Mosquitoes that had bitten dengue patients were taken to a higher region where dengue had never occurred and allowed to bite two healthy persons. Both developed the disease and as they were protected from other mosquitoes until they had recovered, the disease did not spread to others of the community. These and other observations seem to make a complete chain of evidence, and most medical men to-day accept the theory as well proved and in their practice take every precaution to prevent the spread of the disease by keeping the infected patient from being bitten by the mosquitoes.

The yellow fever mosquito is also suspected of carrying this same disease, and it is possible that other species are also concerned. If it is true that the parasite can be carried by several different species of mosquitoes this would account very largely for its rapid spread wherever it is introduced into a community. Where it occurs outside the tropics it is only in the warm summer months when mosquitoes are always abundant.

MALTA OR MEDITERRANEAN FEVER

This is also a tropical and subtropical disease that occasionally gets up into the temperate region, sometimes occurring in the United States. The fever begins with a severe headache, and other symptoms follow. It is usually of the remittent type and may continue for some months.

It is caused by minute bacteria (Micrococcus melitensis) and is a very infectious but not usually contagious disease. The germ is readily conveyed by inoculation, and several investigators have sought to show that the mosquito often serves as the inoculating agent. The disease is especially prevalent during the mosquito season, and has twice been conveyed to monkeys by infected insects.

LEPROSY

This loathsome disease has long been known to be caused by a particular bacillus (Bacillus lepr?, but the way in which this organism gains an entrance into the system is still unknown. Many theories have been propounded, but none of them has been well established. Within recent years the possibility of insects carrying the germ and in one way or another transmitting it to healthy individuals has been suggested and much discussed. As the lepr?bacilli are present in the skin and ulcers of leprous patients, insects sucking the blood or feeding on the sores could not help taking some of them into their body or becoming contaminated. These bacilli have been found at various times in the stomach or intestine of mosquitoes, fleas and bedbugs. So it is believed by some that these and other insects, such as lice and flies, may sometimes transmit the disease. On a previous page we have referred to the possibility of the face-mites acting as disseminators of leprosy.

Leprosy occurs most commonly among people where little attention is paid to bodily cleanliness. Such people are usually freely infested with various parasites that thrive well in the filth, so if the germs can be transmitted in this way the carriers are there in abundance.

The fact that the sores usually occur on exposed parts of the body has been pointed to as evidence that inoculation is due to such insects as flies and mosquitoes. It has been noted that leprosy is frequently very common in regions where elephantiasis occurs, suggesting the possibility of the same carrier, the mosquito, for both diseases. So while there is as yet very little evidence one way or the other, insects that are found around leprous patients are to be regarded with suspicion, for until we know more definitely just how the disease is communicated the insects must be looked on as possible sources of contamination.

KALA-AZAR OR DUM-DUM FEVER

This is a very fatal infectious disease of many tropical and subtropical regions, spreading terror among the natives wherever it occurs. It is caused by the presence in the system of Protozoan parasites, the so-called Leishman-Donovan bodies, that have recently been studied by several observers.

Dr. W.S. Patton of the Indian Medical Service has been making some extensive experiments with the common bedbug of India (Cimex rotundatus) which seem to demonstrate fully that this insect is responsible for the transmission of the parasite that causes the disease. He has found the parasite in all stages of development in the bedbug. This, taken with a number of other observations in regard to the tendency of the disease to cling to particular houses, makes a strong case against the bedbug. Manson, however, believes that the parasite may be transmitted by other agents also, possibly by means of flies that visit the sores or in other ways.

ORIENTAL SORE

This disease, once supposed to be confined to the Orient, is now found to be rather widely distributed throughout the tropics, where it is sometimes very prevalent. It is caused by the presence in the system of a parasite very similar to or identical with the one causing kala-azar and is regarded by some as a modified form of that disease. The patient is affected with one or more serious sores or ulcers which usually occur on exposed parts of the body.

The parasite that causes the disease is supposed to be carried by insects either directly or indirectly.

In the latter case the insect may act as an intermediate host.

Dogs and camels are also attacked by this disease and may be sources of infection.

###

www.ingramcontent.com/pod-product-compliance
Lightning Source LLC
Chambersburg PA
CBHW070910180526
45168CB00005B/1993